Lecture Notes in Mathematics

W9-ABM-7

Edited by A. Dold and B. Eckmann

1020

Non Commutative Harmonic Analysis and Lie Groups

Proceedings of the International Conference
Held in Marseille Luminy, 21–26 June, 1982

Edited by J. Carmona and M. Vergne

Springer-Verlag
Berlin Heidelberg New York Tokyo 1983

Editors

Jaques Carmona
Université d' Aix Marseille II, Departement de Mathématiques de Luminy
70 Route L. Lachamp, 13288 Marseille Cedex 2, France

Michèle Vergne
CNRS Université de Paris VII, UER Mathématiques
2, Place Jussieu, 75221 Paris Cedex 05, France

AMS Subject Classifications (1980): 17B35, 17B45, 22D10, 22E30, 22E46, 43A25

ISBN 3-540-12717-8 Springer-Verlag Berlin Heidelberg New York Tokyo
ISBN 0-387-12717-8 Springer-Verlag New York Heidelberg Berlin Tokyo

Library of Congress Cataloging in Publication Data. Main entry under title: Non commutative harmonic analysis and Lie groups. (Lecture notes in mathematics; 1,020) English and French. Proceedings of the Fifth Colloque "Analyse harmonique non commutative et groupes de Lie," held at Marseille-Luminy, June 21–26, 1982. 1. Harmonic analysis–Congresses. 2. Lie groups–Congresses. I. Carmona, Jacques, 1934-. II. Vergne, Michèle. III. Colloque "Analyse harmonique non commutative et groupes de Lie" (5th: 1982: Marseille France) IV. Series: Lecture notes in mathematics (Springer-Verlag); 1,020. QA3.L28 no. 1,020
510s [512'.55] 83-17111 [QA403]
ISBN 0-387-12717-8 (U.S.)

This work is subject to copyright. All rights are reserved, whether the whole or part of the material is concerned, specifically those of translation, reprinting, re-use of illustrations, broadcasting, reproduction by photocopying machine or similar means, and storage in data banks. Under § 54 of the German Copyright Law where copies are made for other than private use, a fee is payable to "Verwertungsgesellschaft Wort", Munich.

© by Springer-Verlag Berlin Heidelberg 1983
Printed in Germany

Printing and binding: Beltz Offsetdruck, Hemsbach/Bergstr.
2146/3140-543210

MathSci

PREFACE

Le cinquième Colloque "Analyse Harmonique Non Commutative et Groupes de Lie" s'est tenu à Marseille - Luminy du 21 au 26 Juin 1982 dans le cadre du nouveau Centre International de Rencontres Mathématiques.

Les participants noteront que la liste des articles publiés ci-dessous ne correspond pas complètement aux conférences présentées durant le Colloque. C'est le cas, en particulier, pour des travaux dont la publication détaillée était prévue par ailleurs, ou de résultats dûs à plusieurs coauteurs.

Outre les participants à cette rencontre, nous tenons à remercier l'U.E.R. de Marseille - Luminy et le Centre International de Rencontres Mathématiques qui ont rendu possible la tenue de ce Colloque, ainsi que le secrétariat du Laboratoire de Mathématiques qui a assuré la préparation du présent volume.

Jacques CARMONA

Michèle VERGNE

NON COMMUTATIVE HARMONIC ANALYSIS AND LIE GROUPS

L^2 index and unitary representations

M. W. Baldoni Silva

§1. Introduction.

In this paper we investigate the representations that contribute to the index of the Dirac operator. Let G be a connected real semisimple Lie group with finite center, and let K be a maximal compact subgroup of the same rank as G. We assume that $G \subseteq G_{\mathbb{C}}$, where $G_{\mathbb{C}}$ is a linear complexification of G. Let $\mathfrak{g}_0$ and $\mathfrak{k}_0$ be the Lie algebra of G and K respectively. Passing if needed to a suitable double covering of G, we can assume that the isotropy representation $K \to So(\mathfrak{g}_0/\mathfrak{k}_0)$ lift to $Spin(\mathfrak{g}_0/\mathfrak{k}_0)$. Since $\dim(G/K)$ is even, the spin representation (s,S) of $Spin(\mathfrak{g}_0/\mathfrak{k}_0)$ breaks up into two half spin representations s^+,s^-, and correspondingly $S = S^+ \oplus S^-$.

Let η be an irreducible finite dimensional representation of K and let V_η be the corresponding space. If we let $\tilde{\mathcal{E}}^\pm_\eta$ denote respectively the homogeneous vector bundle on G/K defined by the K-modules $E^\pm = V_\eta \otimes S^\pm$ then we may identify the L^2 cross-section of such a bundle as $(L^2(G)\otimes E^\pm)^K$ and we have a corresponding Dirac operator $\tilde{\mathcal{D}}^+ : (L^2(G)\otimes E^+)^K \to (L^2(G)\otimes E^-)^K$ defined as in [P]. Since $\tilde{\mathcal{D}}^+$ is G invariant, it drops down to an elliptic differential operator $\mathcal{D}^+_\Gamma$ on $\Gamma\backslash G/K$ with coefficients in the bundle $\mathcal{E}^\pm = \tilde{\mathcal{E}}^\pm/\Gamma$, Γ being a discrete torsion-free subgroup of G of finite covolume. The L^2 index of $\mathcal{D}^+_\Gamma$ is finite ([M], [A]). More precisely let $\hat{G}$ denote the set of all equivalence classes of irreducible unitary representations π of G and write $m_\Gamma(\pi)$ for the multiplicity of π in $L^2_d(\Gamma\backslash G)$. Then

$$\text{index } \mathcal{D}^+_\Gamma = \sum_{\pi\in\hat{G}} m_\Gamma(\pi)\left(\dim \operatorname{Hom}_K(H(\pi),E^+) - \dim \operatorname{Hom}_K(H(\pi),E^-)\right)$$

$H(\pi)$ being the representation space of π.

The aim of this paper is, fixed a K-module η, to describe all the π's for which $m(\pi,\eta) = \dim \operatorname{Hom}_K(H(\pi),E^+) - \dim \operatorname{Hom}_K(H(\pi),E^-) \neq 0$ and to compute it, when the real rank of G is one.

If π is an irreducible unitary representation with regular integral infinitesimal character, then [V1] gives a necessary condition for $m(\pi,\eta)$ not to be zero. This condition says that π is obtained in terms of parabolic induction via the Vogan-Zuckerman functor and allows one to proceed and compute $m(\pi,\eta)$ as shownin theorem 2.1 below. If the real rank of G is one, this necessary condition is also sufficient in view of the classification of [B-B] of the irreducible unitary representations for real rank one groups. Thus one obtains a complete classification. If π has singular integral infinitesimal character we have to use in some cases a more direct approach.

The new results in this paper are joint work with D. Barbasch.

§2. Notation and main results.

If H is a real Lie group we will write $\mathfrak{h}_0$ for the corresponding Lie algebra and $\mathfrak{h} = (\mathfrak{h}_0)_{\mathbb{C}}$ for the complexification.

Let now G be a Lie group satisfying the same assumption as in the introduction, and let θ be a Cartan involution of $\mathfrak{g}_0$ with Cartan decomposition $\mathfrak{g}_0 = \mathfrak{k}_0 \oplus s_0$. Fix a Cartan subalgebra $\mathfrak{t}_0$ of $\mathfrak{g}_0$ contained in $\mathfrak{k}_0$ and let K and T be the analytic subgroups corresponding to $\mathfrak{k}_0$ and $\mathfrak{t}_0$. Let $\Delta = \Delta(\mathfrak{g},\mathfrak{t})$ be the roots of $\mathfrak{g}$ relative to $\mathfrak{t}$, and let $\Delta(\mathfrak{k}) = \Delta(\mathfrak{k},\mathfrak{t})$ be the corresponding compact roots. If $V \subseteq \mathfrak{g}$ is a $\mathfrak{t}$-invariant subspace, write $\Delta(V)$ for the roots of $\mathfrak{t}$ in V, and $\rho(V) = 1/2\ \Sigma_{\alpha\in\Delta(V)}\ \alpha$.

Fix a system of positive compact roots $\Delta^+(\mathfrak{k})$. Given a θ stable (cf. [V 2]) parabolic subalgebra $\mathfrak{q} = \ell + u$ of $\mathfrak{g}$ with $\ell \supseteq \mathfrak{t}$, we fix a positive system $\Delta^+(\ell \cap \mathfrak{k}) \subseteq \Delta^+(\mathfrak{k})$ for $\Delta(\ell \cap \mathfrak{k},\mathfrak{t})$.

Let L be the subgroup corresponding to $\ell_0 = \ell \cap \mathfrak{g}_0$. Irreducible K types or $L \cap K$ types will be parametrized by the highest weight with respect to $\Delta^+(\mathfrak{k})$ or $\Delta^+(\ell\cap\mathfrak{k})$.

Fix a positive system of roots Δ^+ for Δ such that $2\rho_c = \Sigma_{\alpha\in\Delta^+(\mathfrak{k})}\ \alpha$ is dominant for Δ^+ and normalize s^+ and s^- such that on T', the regular elements of T, the characters of $s^{\pm}$ satisfy:

$$\mathrm{ch}(s^+-s^-)|_{T'} = \prod_{\alpha\in\Delta^+-\Delta^+(\mathfrak{k})} (e^{\alpha/2}-e^{-\alpha/2}).$$

We want to compute

$$m(\pi,\eta) = \dim \mathrm{Hom}_K(H(\pi),V_\eta\otimes S^+) - \dim \mathrm{Hom}_K(H(\pi),V_\eta\otimes S^-)$$

for π a irreducible unitary representation of G and for η a fixed K-module, $H(\pi)$ and $V(\eta)$ being the corresponding representation spaces.

Without loss of generality we may and do exclude the case that π is a discrete series or a nondegenerate limit of discrete series.

Let $\mathrm{ch}\,\pi$ be the distribution character of π, and let $\mathrm{ch}\,\eta$ be the character of η, η being a K-module. Let $D = \Pi_{\alpha\in\Delta^+} (e^{\alpha/2}-e^{-\alpha/2})$. We recall the following facts from [A-S]. If X_π denote the infinitesimal character of π, then

1) $m(\pi,\eta) \neq 0$ only if $X_\pi = X_{\eta+\rho_c}$

2) $m(\pi,\eta) = (-1)^q a_\eta$ where a_η is defined by

$$\mathrm{ch}(s^+-s^-)\mathrm{ch}_\pi|_{T'} = \sum_{\gamma\in\hat{K}} a_\gamma \mathrm{ch}\,\gamma|_{T'} \qquad q = 1/2(\dim \mathfrak{g} - \dim \mathfrak{k})$$

Let Φ be a system of positive roots that makes $\eta + \rho_c$ dominant, and set $\lambda(\eta) = \eta + \rho_c - \rho(\Phi)$, where $\rho(\Phi) = 1/2\,\Sigma_{\alpha\in\Phi}\,\alpha$. We denote by $\mathcal{R}^*$ the Vogan-Zuckerman functor as defined in [V2].

Theorem 2.1. Suppose G has real rank one and $\eta + \rho_c$ is regular. Then the irreducible unitary representations for which $m(\pi,\eta) \neq 0$ are given by the following procedure.

Let $\mathfrak{q} = \ell + u$ be a θ-stable parabolic subalgebra such that $\lambda = \lambda(\eta)$ satisfies $(\lambda,\alpha) = 0$ for $\alpha \in \Delta(\ell)$ and $(\lambda,\alpha) \geq 0$ for $\alpha \in \Delta(u)$. Then $\pi = \mathcal{R}^s \pi^L_\lambda$ and $m(\pi,\eta) = (-1)^{q_L}(-1)^{|-\Delta^+ \cap \Phi|}$, where $q_L = 1/2(\dim \ell - \dim \ell \cap \mathfrak{k})$ and $s = \dim(u \cap \mathfrak{k})$.

Remark. In [V1] it is proved in full generality, i.e., without the rank one assumption, the fact that if $m(\pi,\eta) \neq 0$ then there exists a θ-stable parabolic subalgebra and a λ' as in the theorem so that $\pi = \mathcal{R}^s \pi^L_{\lambda'}$. This is the necessary condition to which we alluded in the introduction.

Proof: By the classification of the irreducible unitary representations of real rank one groups of [B-B] all the representations in the conclusion of the theorem are unitary and conversely all the unitary irreducible representations with regular integral infinitesimal character are obtained in this way.

Therefore it is enough to prove that any representation of the form $\mathcal{R}^s \pi^L_{\lambda'}$ with λ' as in the theorem has $m(\pi,\eta) = 0$ unless $\lambda' = \lambda$ and in this case $m(\pi,\eta) = (-1)^{q_L}(-1)^{|-\Delta^+ \cap \Phi|}$.

Thus suppose that $\mathfrak{q} = \ell + u$ is a θ-stable parabolic subalgebra and $\lambda' \in \hat{T}$ is such that $(\lambda',\alpha) = 0$ for $\alpha \in \Delta(\ell)$ and $(\lambda',\alpha) \geq 0$ for $\alpha \in \Delta(u)$. Let $\Psi(\ell)$ be a positive system for $\Delta(\ell)$ that makes $2\rho(\ell\cap k) = \Sigma_{\alpha\in\Delta^+(\ell\cap k)}\alpha$ dominant, and set $\Psi = \psi(\ell) \cup \Delta(u)$. Then $\pi = \mathcal{R}^s\pi^L_{\lambda'}$ is an irreducible unitary representation by [B-B] and $\mathcal{R}^i\pi^L_{\lambda'} = 0$ for $i < s$ by [V2]. We want to compute $m(\pi,\eta)$ for such a representation.

For any irreducible representation π^L of L we denote by $e(\pi^L)$ the virtual character

$$e(\pi^L) = \sum_i (-1)^i ch(\mathcal{R}^i\pi^L)$$

and refer to it as the Euler characteristic. The Euler characteristic is a well defined map from the ring of virtual characters of L to the ring of virtual characters of G. Furthermore it commutes with coherent continuation. Suppose

$$ch\ \pi^L_{\lambda'} = \Sigma\ c_j\ ch\ \pi^L(\gamma_j)$$

where c_j are integers and $\pi^L(\gamma_j)$ is a discrete series or a principal series representation. Then

$$e(\pi^L_{\lambda'}) = \Sigma\ c_j e(\pi^L(\gamma_j)).$$

We note that $\gamma_j + \rho(u) = w(\lambda'+\rho(\psi))$ with $w \in W(\ell)$, $W(\ell)$ being the Weyl group of ℓ. Since $(\lambda'+\rho(\psi),\alpha) > 0$ for $\alpha \in \Delta(u)$, we obtain $(\gamma_j+\rho(u),\alpha) > 0$ for $\alpha \in \Delta(u)$. It follows from [V 2, theorem 8.2.15] that

$$e(\pi^L(\gamma_j)) = (-1)^s \mathrm{ch}\, \mathcal{R}^s \pi^L(\gamma_j) = (-1)^s \mathrm{ch}\ \pi(\gamma_j+\rho(u)).$$

Thus

$$\mathrm{ch}\ \pi = (-1)^s e(\pi^L_{\lambda'}) = \Sigma\ c_j\ \mathrm{ch}\ \pi(\gamma_j+\rho(u)).$$

Since we need only $\mathrm{ch}\ \pi|_{T'}$ we are interested only in the c_j's for which $\pi^L(\gamma_j)$ is a discrete series representation.

By comparing the expressions for such characters we can write

$$\mathrm{ch}\ \pi^L_{\lambda'}|_{T'} = (-1)^{q_L} \sum_{\substack{w\in W(\ell) \\ w\psi(\ell)\supset\Delta^+(\ell\cap k)}} \mathrm{ch}\ \pi(w\psi(\ell),\ \lambda' + \rho(\ell))|_{T'},$$

where $\rho(\ell) = 1/2\ \Sigma_{\alpha\in\psi(\ell)}\ \alpha$ and where the discrete series on the right side is the one whose chamber is $w\psi(\ell)$ and whose infinitesimal character is $\lambda' + \rho(\ell)$. Then

$$\mathrm{ch}\ \pi|_{T'} = (-1)^{q_L} \sum_{\substack{w\in W(\ell) \\ w\psi(\ell)\supset\Delta^+(\ell\cap k)}} \mathrm{ch}\ \pi(w\psi(\ell)\cup\Delta(u),\ \lambda' + \rho(\psi))|_{T'}$$

Writing out explicitly the expression on the right hand side, one obtains immediately that $m(\pi,\eta) = 0$ unless $w(\lambda'+\rho(\psi)) = \lambda' +$

$w\rho(\psi) = \eta + \rho_c$ for some $w \in W(\ell)$. In this case

$$m(\pi,\eta) = (-1)^{q_L}(-1)^{|-\Delta^+ \cap w\psi(\ell) \cup \Delta(u)|} .$$

Because of the regularity of $\eta + \rho_c$ we see that $m(\pi,\eta) = 0$ unless $\lambda = \lambda'$ and $w\psi(\ell) \cup \Delta(u) = \Phi$. Then $m(\pi,\eta) = (-1)^{q_L}(-1)^{|-\Delta^+ \cap \Phi|}$, and the theorem follows.

We now give an idea of how to compute $m(\pi,\eta)$ when $\eta + \rho_c$ is singular. We omit the details. Since $\eta + \rho_c$ is regular with respect to $\Delta(\mathfrak{k})$, $\eta + \rho_c$ cannot be singular with respect to two adjacent noncompact roots in the Dynkin diagram for the simple roots of Φ. Thus in real rank one, $\eta + \rho_c$ is singular with respect to exactly one simple noncompact root in Φ, say β. Then x_π coincides with the infinitesimal character of a limit of discrete series. We note also that if $w(\eta+\rho_c)$ is dominant with respect to $\Delta^+(\mathfrak{k})$, then $w(\eta+\rho_c)$ must be singular with respect to a compact root unless $w(\eta+\rho_c) = \eta + \rho_c$; hence $w = id$ or the simple reflection about β. Thus

$$\mathrm{ch}\ \pi|_{T'} = (-1)^q m(\eta+\rho_c,\pi) \cdot \frac{\sum_{w\in W(\mathfrak{k})} \varepsilon(w) e^{w(\eta+\rho_c)}}{D}$$

where $W(\mathfrak{k})$ is the Weyl group of $\mathfrak{k}$.

To compute $m(\eta+\rho_c,\pi)$ if real rank of G is one we argue as follows. By the classification in [B-B] we have either that

1) $\pi = \mathcal{R}^s \pi^L_{\lambda'}$, with $\mathfrak{q} = \ell + u$ a θ-stable parabolic and

$$(\lambda',\alpha) = 0, \quad \alpha \in \Delta(\ell), \quad (\lambda'+\rho(\ell)+\rho(u),\alpha) \geq 0 \quad \text{for} \quad \alpha \in \Delta(u)$$

(with equality for at least one $\alpha \in \Delta(u)$) or else

2) $\mathfrak{g} = sp(n,1)$ and π has a lowest K-type of a very special form which can be given explicitly.

In the first situation the argument given in the regular case still works, with obvious modifications, and one gets that $m(\pi,\eta) = 0$ unless there exists $w \in W(\ell)$ such that $w(\lambda'+\rho(\ell)+\rho(u)) = \eta + \rho_c$. (Such w is unique). When w exists then $m(\pi,\eta) = (-1)^{P_L}(-1)^{|-\Delta^+ \cap w\psi(\ell) \cup \Delta(u)|}$.

In the second situation one has to compute directly $D \operatorname{ch} \pi|_{T'}$. This is not very hard because the composition factors of π are known [B-K] and one can proceed to compute $m(\pi,\eta+\rho_c)$.

Departimento di Matematica
Università degli Studi di Trento
38050 Povo (Trento)
Italy

Bibliography

[A] M. F. Atiyah, "Elliptic operators, discrete groups and von Neumann algebras", Astérisque 32/33, (1976), 43-72.

[AS] M. F. Atiyah, W. Schmid, "A geometric construction of the discrete series", Inv. Math., 42(1977), 1-62.

[BB] M. W. Baldoni Silva, D. Barbasch, "The unitary dual for real rank one semisimple Lie groups". Preprint.

[BK] M. W. Baldoni Silva, H. Kraejević, "Composition factors of the principal series representations of the group $Sp(n,1)$", TAMS, 262, (1980), 447-471.

[M] H. Moscovici, "L^2 index of elliptic operators on locally symmetric spaces of finite volume". Preprint.

[P] R. Parthasarathy, "Dirac operators and the discrete series", Ann. of Math., 96, (1972), 1-30.

[V1] D. Vogan, "Unitary representations with cohomology", Preprint.

[V2] D. Vogan, "Representations of real reductive Lie groups", Progress in Mathematics, Birkhauser, 1981.

SUR LA CLASSIFICATION DES MODULES ADMISSIBLES IRREDUCTIBLES

par

Jacques CARMONA

0. Introduction.

0.1. Etant donné un groupe de Lie semi-simple connexe réel G, de centre fini, d'algèbre $\mathfrak{g}$, on fixe un sous-groupe compact maximal K de G, d'algèbre κ , on désigne par θ la conjugaison et par $\mathfrak{g} = \kappa + \mathfrak{p}$ la décomposition de Cartan correspondante. Tout sous-groupe parabolique P de G admet une décomposition de Langlands $P = MAN$ pour laquelle la composante déployée A est un sous-groupe vectoriel dont l'algèbre $\mathfrak{A}$ est contenue dans $\mathfrak{p}$, MA étant le centralisateur de A dans G. Si $\mathfrak{m}$ et $\mathfrak{n}$ sont les algèbres de Lie de M et N respectivement, on note $\Phi(P,A)$ l'ensemble des poids de $\mathfrak{A}$ dans $\mathfrak{n}$ et on définit

(0.1.1) $$\rho_P = \frac{1}{2} \sum_{\alpha \in \Phi(P,A)} \alpha ,$$

chaque poids α figurant avec sa multiplicité. On fixe une fois pour toutes un sous-groupe parabolique minimal $P_\circ = M_\circ A_\circ N_\circ$ de G; on dira que la paire (P,A) est standard si $P \supseteq P_\circ$ et $A \subseteq A_\circ$. Enfin, on note $\mathfrak{g}_{\mathbb{C}}$ la complexifiée de $\mathfrak{g}$ et $\mathcal{E}_{\mathbb{C}}$ le complexifié de tout sous-espace $\mathcal{E}$ de $\mathfrak{g}$.

0.2. Pour tout $(\mathfrak{g},K)$-module admissible de type fini $\mathcal{E}$, l'espace quotient $\mathcal{E}_{\mathfrak{n}} = \mathcal{E}/_{\mathfrak{n}.\mathcal{E}}$ est un $(\mathfrak{m}, M\cap K)$-module admissible de type fini (voir [1] Ch. IV) somme directe des sous-modules

(0.2.1) $$\mathcal{E}_{\mathfrak{n}\,\xi} = \{ v \in \mathcal{E}_{\mathfrak{n}} \ / \ \forall H \in \mathfrak{A} \ \exists k \in N \quad (H - \xi(H)Id)^k . v = 0 \} ,$$

où ξ parcourt le dual complexe $\mathfrak{A}'_{\mathbb{C}}$ de $\mathfrak{A}_{\mathbb{C}}$. On sait qu'à tout élément ξ de

(0.2.2) $$e(P,\mathcal{E}) = \{ \xi \in \mathfrak{A}'_{\mathbb{C}} \ / \ \mathcal{E}_{\mathfrak{n}\,\xi} \neq 0 \} ,$$

et à tout M-module admissible de type fini $(\sigma, \mathcal{H}_\sigma)$ dont l'espace des vecteurs $M\cap K$-finis $(\mathcal{H}_\sigma)_\circ$ est isomorphe à un quotient non nul de $\mathcal{E}_{\mathfrak{n}\xi}$, est associé un $(\mathfrak{m},K)$-morphisme non nul $\mathcal{E} \to \mathfrak{I}_{P,\sigma,\nu}$ où $\nu = \xi - \rho_P$ et $\mathfrak{I}_{P,\sigma,\nu}$ désigne l'espace de la série principale induite de P à G par le couple (σ,ν) , (voir [1] Ch.IV).

0.3. Pour établir sa classification, Langlands utilise une construction géométrique permettant d'associer à tout $(\mathfrak{g},K)$-module irréductible $\mathcal{E}$, par l'intermédiaire d'un élément extrémal de $e(P_\circ,\mathcal{E})$, une paire parabolique standard (P,A), une représentation tempére $(\sigma,\mathcal{H}_\sigma)$ de M et un caractère complexe strictement $\Phi(P,\mathfrak{A})$-dominant ν de $\mathfrak{A}$, de telle sorte que ξ soit isomorphe à un quotient de $\mathfrak{I}_{P,\sigma,\nu}$. Dans la première partie de ce travail, nous proposons une interprétation de cette construction qui nous permet de simplifier notablement l'exposé de Borel-Wallach ([1] Ch.IV).

0.4. Soit B un sous-groupe de Cartan fondamental θ-stable de G, $\mathfrak{b}$ son algèbre de Lie, $\mathfrak{t} = \mathfrak{b} \cap \kappa$ et Δ (respmt. Δ_κ) le système des racines de $(\mathfrak{g},\mathfrak{b})$, (respmt. $(\kappa,\mathfrak{t})$). On fixe un système Δ_κ^+ de racines positives pour Δ_κ et on définit:

(0.4.1) $$\rho_c = \frac{1}{2} \sum_{\nu\in\Delta_\kappa^+} \nu .$$

Toute classe de représentation irréductible de dimension finie (appelée κ-type) de κ est alors caractérisée par son poids Δ_κ^+-dominant μ.

0.5.Etant donné un κ-type μ, on choisit un système Δ^+ de racines positives pour Δ, θ-stable et rendant $\mu+2\rho_c$ Δ^+- dominant. Si

(0.5.1) $$\lambda_\mu = \mu + 2\rho_c - \rho$$

où

(0.5.2) $$\rho = \frac{1}{2} \sum_{\alpha\in\Delta^+} \alpha ,$$

Vogan ([10] Prop. 1.4) démontre qu'il existe une décomposition

(0.5.3) $$\lambda_\mu = (\lambda_\mu)_\circ - \sum_{j=1}^{p} c_j \beta_j ,$$

où $(\lambda_\mu)_\circ$ est Δ^+-dominant, $c_j \geq 0$, $(1\leq j\leq p)$, et $\{\beta_1,\ldots,\beta_p\}$ un système orthogonal de racines imaginaires de Δ^+, orthogonales à $(\lambda_\mu)_\circ$. En outre, $(\lambda_\mu)_\circ$ ne dépend que de μ et non du choix de Δ^+. Dans la deuxième partie de ce travail, nous proposons une approche différente de ces résultats. Notre méthode nous permet de simplifier notablement les démonstrations de Vogan et, notamment, de trivialiser la démonstration de l'unicité de $(\lambda_\mu)_\circ$, (Prop. 2.2).

0.6. Il est remarquable de constater que les deux classifications dont nous disposons sont basées sur la même construction géométrique: la projection d'une forme "extrémale" sur le cône convexe fermé des poids dominants. On peut espérer que cette présentation plus simple facilitera la comparaison des deux classifications, en permettant par exemple de préciser comment les informations obtenues par l'une ou l'autre de ces méthodes se complètent. Signalons que, dans l'un et l'autre cas, le caractère infinitésimal peut être défini par une forme linéaire dont la restriction à un sous-espace convenable (dépendant de la projection ci-dessus) coïncide avec cette projection ,(Vogan [10] Prop.5.8 et Harish Chandra [4] p.161) .

1. La construction de Langlands.

1.1. Dans tout ce qui suit, on désigne par V un espace vectoriel réel de dimension finie, muni d'un produit scalaire euclidien $(u,v) \to \langle u,v\rangle$, et par $u \to \|u\| = (\langle u,u\rangle)^{1/2}$ la norme euclidienne correspondante. On fixe un cône ouvert convexe $\mathcal{C}$ de V , on note

(1.1.1) $\overset{\circ}{\mathcal{C}} = \{ v\in V \ / \ \forall w\in\mathcal{C} \ \ \langle v,w\rangle > 0 \}$

le cône dual, $\overline{\mathcal{C}}$ et $\overline{\mathcal{C}}^\circ$ les adhérences respectives de $\mathcal{C}$ et $\mathcal{C}^\circ$.

1.2. Proposition.

Pour tout couple $(v,v_\circ)$ de $V\times\overline{\mathcal{C}}$, les conditions suivantes sont équivalentes.

(i) Pour tout élément w de $\overline{\mathcal{C}}$

(1.2.1) $\|v-v_\circ\| \leq \|v-w\|$;

(ii) Pour tout élément w de $\overline{\mathcal{C}}$

(1.2.2) $\langle v-v_\circ,w-v_\circ\rangle \leq 0$;

(iii) Le vecteur $v_\circ-v$ appartient à $\overline{\mathcal{C}}^\circ$ et

(1.2.3) $\langle v-v_\circ,v_\circ\rangle = 0$.

En outre, pour tout élément v de V , il existe un élément $v_\circ$ de $\overline{\mathcal{C}}$ et un seul, pour lequel le couple $(v,v_\circ)$ vérifie les conditions équivalentes ci-dessus.

Démonstration

Ce résultat, relatif à la projection sur un convexe fermé dans un espace de Hilbert, est classique. Précisons seulement l'équivalence de (ii) et (iii). (ii) $\Rightarrow$(iii).Si on applique (1.2.2) aux vecteurs $w=\frac{1}{2}v_\circ$ et $w=2v_\circ$, on obtient deux inégalités qui impliquent (1.2.3). La relation (1.2.2) exprime alorsl'appartenance de $v_\circ-v$ au cône fermé dual $\overline{\mathcal{C}}^\circ$ de $\overline{\mathcal{C}}$. La réciproque est évidente.

1.3. On fixe une base réelle $\{\alpha_1,\ldots,\alpha_n\}$ de V et on désigne par $\{\omega_1,\ldots,\omega_n\}$ la base duale, c'est à dire la base de V définie par les relations

(1.3.1) $$\langle\omega_i,\alpha_j\rangle = \delta_{ij} \quad , \quad (1\leq i,j\leq n) \quad ,$$

où δ_{ij} désigne le symbole de Krönecker. On suppose désormais que

(1.3.2) $$\mathbb{C} = \{v\in V \ / \ v=\sum_{j=1}^{n} t_j\omega_j \ , \ (t_1,\ldots,t_n)\in (\mathbb{R}_+^*)^n \} \quad ,$$

où $\mathbb{R}_+$ (respmt. $\mathbb{R}_+^*$) désigne l'ensemble des réels positifs (respmt. strictement positifs) . Dans ce cas

(1.3.3) $$\mathbb{C}^\circ = \{v\in V \ / \ v=\sum_{j=1}^{n} s_j\alpha_j \ , \ (s_1,\ldots,s_n)\in (\mathbb{R}_+^*)^n \} \quad .$$

<u>1.4.Corollaire</u> (Langlands, voir [1]).

<u>Les notations et les hypothèses sont celles de 1.3</u> . <u>Pour tout vecteur v de V</u> , <u>il existe une partie unique $F(v)$ de $[1,n]$ telle que:</u>

(1.4.1) $$v = \sum_{j\notin F(v)} t_j\omega_j - \sum_{i\in F(v)} s_i\alpha_i \quad ,$$

<u>avec</u> $t_j > 0$, $s_i \geq 0$, <u>($j\notin F(v)$, $i\in F(v)$)</u>.
<u>Dans ce cas:</u>

(1.4.2) $$v_\circ = \sum_{j\notin F(v)} t_j\omega_j \quad .$$

Démonstration.
A tout élément $v_\circ$ de $\overline{\mathbb{C}}$, la relation

$$v_\circ = \sum_{j\notin F(v_\circ)} t_j\omega_j \quad ,$$

avec $t_j>0$ si $j\notin F(v_\circ)$, associe une partie $F(v_\circ)$ de $[1,n]$. Si on écrit

$$v_\circ - v = \sum_{i=1}^{n} s_i\alpha_i \quad ,(s_i\in R \ , \ 1\leq i\leq n) \quad ,$$

l'appartenance de $v_\circ - v$ à $\overline{\mathbb{C}}^\circ$ se traduira par

$$s_i \geq 0 \quad , \quad (1\leq i\leq n) \quad ,$$

et, dans ce cas, $v_\circ - v$ sera orthogonal à $v_\circ$ si et seulement si

$$s_j = 0 \quad , \quad (j \notin F(v_\circ)) \quad .$$

Le résultat découle alors de (1.2.(ii)) ; on définit :

$$F(v) = F(v_\circ) \quad .$$

1.5. Définition.

On dira que la base $\{\alpha_1,\ldots,\alpha_n\}$ de V est une base de Langlands si:

(1.5.1) $$<\alpha_i,\alpha_j> \leq 0 \qquad ,(\ 1\leq i\neq j\leq n\ ) \quad .$$

On suppose désormais que $\{\alpha_1,\ldots,\alpha_n\}$ est une base de Langlands de V .

1.6.Corollaire.

Les notations et les hypothèses sont celles de 1.4, 1.5 . Pour tout vecteur v de V , et tout entier j de $[1,n]$:

(1.6.1) $$F(v) = \{\ i\in[1,n]\ /\ \ <\alpha_i,v_\circ> = 0\ \} \quad ;$$

(1.6.2) $$F(v) \subseteq \{\ i\in[1,n]\ /\ \ <\alpha_i,v> \leq 0\ \} \quad .$$

Démonstration.

Si $j\notin F(v)$, on a, avec les notations de (1.4.1):

$$\begin{aligned}<v,\alpha_j> &= <v_\circ,\alpha_j> - \textstyle\sum_{i\in F(v)} s_i <\alpha_i,\alpha_j> \\ &\leq <v_\circ,\alpha_j> = t_j > 0 \quad .\end{aligned}$$

1.7. Remarque.

Si les éléments de F(v) sont deux à deux orthogonaux, l'inclusion (1.6.2) est une égalité .

1.8. A toute partie F de [1,n] , on associe le sous-espace:

(1.8.1) $$V_F = \sum_{j\notin F} \mathbb{R}\omega_j \quad ,$$

(avec $V_F=\{0\}$ si $F=[1,n]$) , et son orthogonal

(1.8.2) $$V_F^{\perp} = \sum_{i\in F} \mathbb{R}\,\alpha_i \quad .$$

On note $v \to v_F$ la projection orthogonale de V sur V_F.

1.9. Proposition.

Les notations et les hypothèses sont celles de 1.6 . Pour toute partie F de [1,n] et tout vecteur v de V vérifiant:

(1.9.1) $$<v,\alpha_i> \leq 0 \qquad ,\ (\ i\in F\ )\ ,$$

on a:

(1.9.2) $$v_F = v + \sum_{i\in F} a_i \alpha_i \quad ,$$

avec $a_i \geq 0$, ($i\in F$) .

Démonstration

Si, dans (1.9.2), l'ensemble

$$A = \{ i\in F \ / \ a_i < 0 \} \quad ,$$

est non vide, écrivons:

$$v_F - \sum_{k\in A} a_k\alpha_k = v + \sum_{i\in F-A} a_i\alpha_i \quad ,$$

$$- \| \sum_{k\in A} a_k\alpha_k \|^2 = < \sum_{k\in A} a_k\alpha_k , v_F - \sum_{k\in A} a_k\alpha_k > \quad ,$$

$$= < \sum_{k\in A} a_k \alpha_k , v> + \sum_{k\in A} \sum_{i\in F-A} a_k a_i <\alpha_k,\alpha_i>$$

$$\geq 0 \quad .$$

C'est exclu.

1.10. Corollaire.

Les notations et les hypothèses sont celles de 1.6 . Il existe une famille a_{ij} ,($1\leq i,j\leq n$), de réels positifs tels que:

(1.10.1) $$\omega_j = \sum_{i=1}^{n} a_{ij} \alpha_i \quad , \ (1\leq j\leq n) .$$

Démonstration

C'est le cas $v=-\omega_j$, $F=[1,n]$, ($1\leq j\leq n$) .

1.11. On munit V de la structure d'espace vectoriel ordonné par le cône $\overline{C}^\circ$, c'est à dire que, pour (v,w) appartenant à $V\times V$ on écrit $v\leq w$ si et seulement si

$$w-v \in \overline{C}^\circ \quad .$$

1.12. Corollaire (Langlands)

Les notations et hypothèses sont celles de 1.6 et 1.11 . Quels que soient les vecteurs v et w de V, on a l'implication

$$v \geq w \Rightarrow v_\circ \geq w_\circ \quad .$$

Démonstration.

On note $F = F(w)$, $u = w_o - v_o$ et

$$u = u_o - \Sigma_{j \in F(u)} \, s_j \alpha_j \qquad , \, (s_j \geqslant 0 \, , \, j \in F(u))$$

Si $j \in F$

$$< \alpha_j , u > = - < \alpha_j , v_o > \leqslant 0 \quad ,$$

et donc, d'après 1.6 $F \subset F(u)$. Supposons que $F(u)$ est distinct de $[1,n]$ et fixons $j \notin F(u)$. Puisque $j \notin F$:

$$0 < < u_o , \omega_j > = < u , \omega_j > = < w_o , \omega_j > - < v_o , \omega_j > = < w , \omega_j > - < v_o , \omega_j >$$
$$\leqslant < w , \omega_j > - < v , \omega_j > \leqslant 0 \quad ,$$

car $v - w$ appartient à $\overset{\circ}{\overline{c}}$. C'est exclu. On a donc $F(u) = [1,n]$, c'est-à-dire le résultat cherché.

1.13. Reprenons les notations et les hypothèses de 0.1-0.2-0.3 . La restriction à $\mathcal{A}_o \times \mathcal{A}_o$ de la forme de Killing de $\mathcal{G}$ définit sur $\mathcal{A}_o$ une structure euclidienne que l'on transporte sur son dual $V = \mathcal{A}_o'$. On note $\{\alpha_1,\ldots,\alpha_n\} \subseteq V$ le système des racines $\Phi(P_o,A_o)$-simples de $(\mathcal{G},\mathcal{A}_o)$. Le cône $\overline{C}$ est alors le cône des poids $\Phi(P_o,A_o)$ dominants . Toute partie F de $[1,n]$ définit un système parabolique de racines restreintes . Si $P_F = M_F A_F N_F$ est le parabolique standard ainsi défini, le parabolique minimal $^*P = P \cap M_F$ de M_F s'écrit $^*P = M_o {}^*A_F {}^*N_F$ où $^*A_F = A_o \cap M_F$ et $^*N_F = N_o \cap M_F$ (voir [1] Ch.IV pour des détails plus précis et les notations).

1.14. Proposition.

Etant donné un $(\mathcal{G},K)$-module admissible de longueur finie $\mathscr{E}$, on fixe un élément ν de $e(P_o,\mathscr{E})$ pour lequel, si $\varphi = - \nu + \rho_{P_o}$, $(Re \, \varphi)_o$ est minimal (voir 1.13). Soient $F = F(Re \, \varphi)$, $\xi = \nu_{|\mathcal{A}_F}$, $(\sigma \mathcal{H}_\sigma)$ un M_F-module admissible de type fini dont l'espace des vecteurs $M_F \cap K$-finis est équivalent à un quotient non nul de $\mathscr{E}_{\eta_F, \xi}$, (voir 0.2). Alors:

a/ $Hom_{(\mathcal{G},K)}(\mathscr{E} , \mathfrak{I}_{P_F, \sigma\xi - \rho_{P_F}}) \neq 0$,(voir 0.3) .

b/ Pour tout $\alpha \in \Phi(P_F,A_F)$

(1.14.1) $\qquad\qquad < Re \, \xi - \rho_{P_F} , \alpha > \, < 0$.

c/ Pour tout $\mu \in e(^*P ,(\mathcal{H}_\sigma)_o)$:

(1.14.2) $\qquad\qquad Re \, \mu - \rho_{*P} \, \geqslant \, 0$.

Démonstration.

On note $P=P_F$, $A=A_F$, ${}^*P={}^*P_F$,..., $\rho_\circ=\rho_{P_\circ}$, $\rho = \rho_P$, ${}^*\rho=\rho^*_P$, etc. et

$$\operatorname{Re}\varphi = \sum_{j\notin F} t_j\,\omega_j - \sum_{i\in F} s_i\,\alpha_i \qquad (t_j>0\ ,\ s_i\geq 0)\ ,$$

de telle sorte que

$$\operatorname{Re}\xi+\rho = (\operatorname{Re}\varphi)_\circ = \sum_{j\notin F} t_j\,\omega_j \quad .$$

On a identifié le dual de $\mathcal{A}$ (respmt. de ${}^*\mathcal{A}$) à l'orthogonal de ${}^*\mathcal{A}$ (respmt de $\mathcal{A}$) dans le dual de $\mathcal{A}_\circ$.

a/ est un résultat standard (voir [1]) et la vérification de b/ est triviale.Pour vérifier c/ , on écrit:

$$\operatorname{Re}\mu - {}^*\rho = \sum_{i\in F} x_i\,\alpha_i \quad .$$

Il s'agit de démontrer que l'ensemble

$$F_1 = \{\ i\in F \ /\ x_i<0\ \} \quad ,$$

est vide. Si $F_2=F-F_1$ et $-\psi = \mu+\xi-\rho_\circ$, on peut écrire:

$$\operatorname{Re}\psi = \sum_{j\notin F} t_j\,\omega_j - \sum_{i\in F} x_i\,\alpha_i$$

$$\leq \sum_{j\notin F} t_j\,\omega_j - \sum_{i\in F_2} x_i\,\alpha_i = \eta \quad .$$

On déduit de 1.14 et 1.4

$$(\operatorname{Re}\psi)_\circ \leq \eta_\circ = \sum_{j\notin F} t_j\,\omega_j = (\operatorname{Re}\varphi)_\circ \quad .$$

Cela signifie, d'après le choix de φ , que

$$(\operatorname{Re}\psi)_\circ = (\operatorname{Re}\varphi)_\circ \quad ,$$

ce qui permet d'écrire

$$\operatorname{Re}\psi = \sum_{j\notin F} t_j\,\omega_j - \sum_{i\in F} y_i\,\alpha_i \quad , \ (y_i\geq 0);$$

Si F_1 est nonvide, on obtient, pour tout $i\in F_1$

$$0 < -x_i = \langle \operatorname{Re}\psi - \eta, \omega_i\rangle = -y_i \leq 0 \quad .$$

1.15. Remarque.

On sait, (voir [1] 3.7 p.124) , que (1.16.2) caractérise les représentations tempérées, et que, pour $\mathcal{E}$ irréductible, le système (P_F,σ,ξ) vérifiant les hypothèses de 1.16 est unique ([1] p.131). On définit ainsi le système des paramètres de Langlands de $\mathcal{E}$.

2. La construction de Vogan.

2.1. On reprend les notations et les hypothèses de 1.1-1.3 dans le contexte suivant. On suppose que V est engendré par un système de racines Δ , on dési-par $\mathcal{W}$ le groupe de Weyl correspondant, et on fixe sur V une structure euclidienne $\mathcal{W}$-invariante (voir [2]). Pour toute racine α de Δ , on notera s_α la symétrie de Weyl de V associée à α. Enfin, on suppose donnée une application $\mathcal{W}$-invariante $\alpha \to n(\alpha)$ de Δ dans l'ensemble $\mathbb{N}^*$ des entiers strictement positifs. A toute chambre de Weyl $\mathbb{C}$ de V est associé un système $\Delta(\mathbb{C})$ de racines positives pour Δ et un vecteur

(2.1.1) $$\rho_{\mathbb{C}} = \frac{1}{2}\sum_{\alpha\in\Delta(\mathbb{C})} n(\alpha)\alpha \quad ,$$ tel que,

(2.1.2) $$2\,\frac{<\rho\ ,\alpha>}{<\alpha,\alpha>} = n(\alpha) > 0 \quad , (\alpha\in\Delta(\mathbb{C})) .$$

2.2. Proposition.

Etant donnés un vecteur v de V et une chambre de Weyl $\mathbb{C}$ dont l'adhérence contient v, la projection $\hat{v}$ de $v-\rho$ sur $\overline{\mathbb{C}}$ ne dépend pas de la Chambre de Weyl $\mathbb{C}$ telle que $v\in\overline{\mathbb{C}}$. De plus:

(2.2.1) $$v-\hat{v} \in \overline{\mathbb{C}} \cap (\rho-\overline{\mathbb{C}}^{\circ})$$

Démonstration.

Soient $\rho=\rho_{\mathbb{C}}$, $F=F(v-\rho)$, et $\{\alpha_1,\ldots,\alpha_n\}$ le système des racines $\Delta(\mathbb{C})$-simples de Δ, de telle sorte que (voir 1.4) :

(2.2.2) $$v-\rho = \sum_{j\notin F} y_j\,\omega_j - \sum_{i\in F} x_i\,\alpha_i \quad ,$$

avec $y_j>0$, $x_i\geq 0$,$(j\notin F, i\in F)$. Cela signifie que:

(2.2.3) $$\hat{v} = \sum_{j\notin F} y_j\,\omega_j$$

Définissons

$$F' = \{\ j\in[1,n] \ / \ <v,\alpha_j> =0\ \} \quad ,$$

et notons $\mathcal{W}'$ le sous-groupe de $\mathcal{W}$ engendré par les symétries s_{α_j} $(j\in F')$. Remarquons que, d'après 1.6

(2.2.4) $$F'\subseteq F \qquad \text{et} \qquad s(\hat{v}) = \hat{v} \quad ,(s\in\mathcal{W}') .$$

D'autre part, si $\mathbb{C}'$ est une autre chambre de Weyl dont l'adhérence contient v, il existe, (voir [2] p.75), un élément s de $\mathcal{W}'$ tel que

$$\mathbb{C}' = s(\mathbb{C}) \quad .$$

Appliquons s aux deux membres de (2.2.2)

$$v-\rho_{\mathbb{C}'} = s(v-\rho) = s(\hat{v} - \sum_{i\in F} x_i \alpha_i)$$

$$= \hat{v} - \sum_{i\in F} x_i s(\alpha_i) \quad .$$

Le système des racines $\Delta(\mathbb{C}')$-simples est $\{s(\alpha_1),\dots,s(\alpha_n)\}$, et, pour $1\leq j\leq n$:

$$<\hat{v},s(\alpha_j)> = <s^{-1}(\hat{v}),\alpha_j> = <\hat{v},\alpha_j> \geq 0 \quad ,$$

avec égalité si $j\in F$. Cela signifie que $\hat{v}$ est la projection de $v-\rho$, sur $\overline{\mathbb{C}}'$. Pour vérifier (2.2.1) , remarquons que (voir (2.1.2) et 1.5):

$$<v-\hat{v},\alpha_j> = <\rho,\alpha_j> - \sum_{i\in F} x_i<\alpha_i,\alpha_j> > 0 \quad ,(j\notin F),$$

et puisque v appartient à $\overline{\mathbb{C}}$,(voir (1.6.1)):

$$<v-\hat{v},\alpha_j> = <v,\alpha_j> \geq 0 \quad , (j\in F)$$

2.3. On suppose désormais que le système Δ de racines est réduit, ce qui entraîne en particulier que $n(\alpha)=1$ pour $\alpha\in\Delta$. On fixe une chambre de Weyl $\mathbb{C}$ et on note $\Delta^+=\Delta(\mathbb{C})$, $\rho=\rho_{\mathbb{C}}$. Pour toute racine β de Δ^+ , on note $l(\beta)$ le nombre de racines Δ^+-simples figurant dans l'expression de β comme somme de racines Δ^+-simples. Pour toute partie S de Δ^+ , on note Δ_S le système des racines de Δ orthogonales aux éléments de S , et $\Delta_S^+ = \Delta_S\cap\Delta^+$. Dans le cas où $S=\{\beta\}$, on écrira $\Delta_\beta = \Delta_S$ et $\Delta_\beta^+= \Delta_S^+$.

2.4. Proposition.

Si β est Δ^+-simple, et si α est Δ_β^+-simple, on est dans l'un des cas suivants.

a/ α est Δ^+-simple.

b/ Il existe une racine Δ^+-simple γ telle que:

(2.4.1) $\quad <\gamma,\gamma> \neq <\beta,\beta> \quad ,$

(2.4.2) $\quad <\gamma,\alpha> > 0 \quad .$

Dans ce cas, $\Delta\cap(R\beta+R\gamma)$ est un système de racines de rang deux (de type B_2 ou G_2) contenant α.

c/ Toute racine Δ^+-simple γ vérifiant (2.4.2) est telle que:

(2.4.3) $<\gamma,\gamma> = <\beta,\beta>$.

Dans ce cas, il existe une suite $j \to \gamma_j$,($\gamma_o=\beta$, $0\leq j\leq p$), de racines Δ^+-simples engendrant un système de type A_{p+1} , avec $\delta_j=\gamma_j+\gamma_{j-1} \in\Delta$, $(1\leq j\leq p)$, et telle que si $s_j=s_{\delta_j}$, $w_j=s_j\ldots s_1$, $(1\leq j\leq p)$, on ait (avec $w_o=1$):

(i) $w_j\alpha$ est $\Delta^+_{\{\gamma_o,\ldots,\gamma_j\}}$-simple , $(0\leq j\leq p)$

(ii) $l(w_j\alpha) < l(w_{j-1}\alpha)$, $(1\leq j\leq p)$.

(iii) $w_p\alpha$ vérifie les hypothèses de a/ ou b/.

Démonstration.

On peut se placer dans le cas où Δ est irréductible, de type différent de G_2 et supposer que α n'est pas Δ^+-simple. Dans ce cas, si γ et δ sont deux racines de Δ telles que

$$<\gamma,\gamma> \geq <\delta,\delta> ,$$

on a

$$<\gamma,\gamma> = m <\delta,\delta> , \quad (m=1 \text{ ou } 2).$$

Soit γ une racine simple vérifiant (2.4.2) de telle sorte que:

(2.4.4) $\alpha' = \alpha-\gamma \in\Delta^+$,

(2.4.5) $<\beta,\gamma> \leq 0$.

Si γ est orthogonale à β , $\alpha = \alpha'+\gamma$ n'est pas Δ^+_β-simple. On a donc:

(2.4.6) $<\beta,\gamma> < 0$,

c'est à dire que $\delta = \beta+\gamma$ appartient à Δ^+ , avec

(2.4.7) $<\beta,\alpha'> = <\beta,\alpha-\gamma> = -<\beta,\gamma> > 0$.

On est donc dans l'un des deux cas suivants.

- Soit $\alpha' = \beta$ et $\alpha = \alpha'+\gamma = \beta+\gamma$ appartient au système de racines de rang deux engendré par β et γ. Comme α est orthogonale à β, ce système est de type B_2 et γ est sa racine longue.

- Soit $\alpha'' = \alpha'-\beta = \alpha-\delta$ est une racine (nécessairement positive)de Δ, ce que nous supposerons désormais.

Si, en outre, (2.4.1) est vérifié, on est dans l'un des deux cas suivants.

- Soit $\gamma+2\beta$ est racine, ce qui implique (voir [2] p. 150), que $\delta= \gamma+\beta$ est orthogonale à β ; $\alpha = \alpha''+\delta$ n'est pas Δ^+_β-simple.

- Soit $\delta' = \beta+2\gamma \in\Delta$, ce qui implique: $<\beta,\delta'> = 0$. Dans ce cas

$$<\alpha,\delta'> \;=\; 2<\alpha,\gamma> \;>0 \qquad ,$$

et on a alors deux possibilités.

- Soit $\alpha = \delta' = \beta+2\gamma$ appartient au système de racines de type B_2 engendré par β et γ, et dont β est la racine longue; c'est le cas b/ .

- Soit $\alpha-\delta' = \alpha-\delta-\gamma = \alpha''-\gamma \subset \Delta^+$ et $\alpha = \delta'+(\alpha-\delta')$ n'est pas Δ^+_β-simple.

Cela signifie que si l'on n'est ni dans le cas a/ , ni dans le cas b/, β et γ engendrent un système de type A_2 et

$$\alpha' = \alpha-\gamma \qquad ,$$
$$\alpha'' = \alpha-\delta \qquad ,$$

sont des racines positives de Δ. Dans ce cas $s_\delta = s_\beta s_\gamma s_\beta$ vérifie (voir [2] p. 158)

(2.4.8) $$(-s_\delta\Delta^+)\cap\Delta^+ = \{\beta,\gamma,\delta\} \qquad ,$$

et donc

(2.4.9) $$s_\delta\Delta^+_\beta = \Delta^+_\gamma \qquad .$$

La racine $s_\delta\alpha$ est donc Δ^+_γ-simple et vérifie:

$$l(s_\delta\alpha) < l(\alpha) \qquad .$$

On conclut par récurrence sur $l(\alpha)$.

2.5. Lemme.

On suppose que Δ est irréductible et réduit, et on désigne par $\mathcal{F}$ l'ensemble des éléments de $\overline{C}°$ qui sont combinaison linéaire à coefficients positifs de sommes de racines Δ^+-simples, adjacentes, et de même longueur. On est alors dans l'un au moins des cas suivants.

a/ Soit Δ est de type A_n ,$(n\geq 2)$, D_{2n+1} ,$(n\geq 2)$, ou E_6 . Dans ce cas, pour tout élément v de $C°$, il existe un élément w de $\mathcal{F}$ tel que:

(2.5.1) $$<v,w> \;>\; 0 \qquad .$$

b/ Le groupe de Weyl $\mathfrak{W}$ de Δ contient l'application (-Id) de V.

Démonstration.

Le fait que l'on soit dans l'un au moins des cas a/ ou b/ résulte [6] Lemme1.1

Vérifions (2.5.1) pour

(2.5.2) $$v = \sum_{j=1}^{n} c_j\,\alpha_j \qquad , \quad (c_j>0 \;\; , \; 1\leq j\leq n) \;.$$

Si Δ est de type A_n, et si on indexe le système $\{\alpha_1,\ldots,\alpha_n\}$ des racines Δ^+-simples de telle sorte que

(2.5.2) $$\langle\alpha_i,\alpha_{i+1}\rangle \neq 0 \quad , \quad (1\leq i\leq n-1) \quad ,$$

on peut choisir

$$w = \begin{cases} (\alpha_1+\alpha_2)+(\alpha_3+\alpha_4)+\ldots+(\alpha_{2p-1}+\alpha_{2p}) & \text{si } n = 2p \\ (\alpha_1+\alpha_2)+(\alpha_2+\alpha_3)+(\alpha_4+\alpha_5)+\ldots+(\alpha_{2p}+\alpha_{2p+1}) & \text{si } n = 2p+1 \end{cases} .$$

Si Δ est de type E_6, et si on indexe le système $\{\alpha_1,\ldots,\alpha_6\}$ des racines Δ^+-simples de telle sorte

(2.5.3) $$\langle\alpha_1,\alpha_2\rangle = \langle\alpha_2,\alpha_3\rangle = \langle\alpha_3,\alpha_4\rangle = \langle\alpha_3,\alpha_5\rangle = \langle\alpha_5,\alpha_6\rangle \neq 0 \quad ,$$

et $c_2 \geq c_5$, on peut choisir:

$$w = \begin{cases} (\alpha_3+\alpha_4)+(\alpha_5+\alpha_6) & \text{si } c_2 \leq c_4 \ , \\ (\alpha_1+\alpha_2)+(\alpha_2+\alpha_3)+(\alpha_3+\alpha_5)+(\alpha_5+\alpha_6) & \text{si } c_4 < c_2 \ . \end{cases}$$

Enfin, si Δ est de type D_{2p+1} et si on indexe le système $\{\alpha_1,\ldots,\alpha_{2p+1}\}$ des racines Δ^+-simples de telle sorte que:

(2.5.4) $$\langle\alpha_i,\alpha_{i+1}\rangle = \langle\alpha_{2p-1},\alpha_{2p+1}\rangle \neq 0 \quad , \quad (1\leq i\leq 2p-1) \quad ,$$

avec $c_{2p} > c_{2p+1}$, on peut choisir:

$$w = (\alpha_1+\alpha_2)+(\alpha_3+\alpha_4)+\ldots+(\alpha_{2p-1}+\alpha_{2p}).$$

2.6. On reprend les notations et les hypothèses de 0.4 . Tout élément α du système Δ des racines de $(\mathfrak{G},\mathfrak{B})$ appartient au dual réel $\mathfrak{B}'_{\mathbb{R}}$ de

(2.6.1) $$\mathfrak{B}_{\mathbb{R}} = \mathfrak{t}_{\mathbb{R}}+\mathfrak{B}_p \quad , \quad (\mathfrak{B}_p=\mathfrak{B}\cap\mathfrak{p},\ \mathfrak{t}_{\mathbb{R}}=\sqrt{-1}\,\mathfrak{t}) \ ,$$

et s'écrit:

(2.6.2) $$\alpha = \alpha_k+\alpha_p$$

où α_k (respmt. α_p) est un élément du dual de $\mathfrak{t}_{\mathbb{R}}$ nul sur $\mathfrak{B}_p$, (respmt; du dual de $\mathfrak{B}_p$ nul sur $\mathfrak{t}_{\mathbb{R}}$). Le fait que $\mathfrak{B}$ soit fondamentale se traduit par:

(2.6.3) $$\alpha_k \neq 0 \quad , \quad (\alpha\in\Delta) \ .$$

En particulier, pour toute racine α , $\alpha - \theta\alpha$ n'est pas racine. La racine est imaginaire (respmt. complexe) si $\alpha_p=0$ (respmt. $\alpha_p\neq 0$); α est imaginaire compacte ou non compacte selon que le sous-espace de poids correspondant est contenu dans $\kappa_{\mathbb{C}}$ ou $\mathfrak{p}_{\mathbb{C}}$. On note Δ_n (respmt. Δ_c) l'ensemble des racines imaginaires non compactes (respmt. des racines complexes) de Δ . Le sytème Δ_k des racines de $(\kappa,\mathfrak{t})$ s'identifie alors à l'ensembles des restrictions à $\mathfrak{t}$ des éléments de $\Delta-\Delta_n$.

2.7. On fixe un système Δ_k^+ de racines positives pour Δ_k; toute classe d'équivalence de représentation irréductible de dimension finie de κ est caractérisée par son poids Δ_k^+-dominant μ . Dans tout ce qui suit, μ est un élément Δ_k^+-dominant de $\mathfrak{t}'_{\mathbb{R}}$. On fixe un système Δ^+ de racines positives pour Δ rendant (voir 0.4) $\mu+2\rho_c$ Δ^+-dominant, et on note:

(2.7.1) $$\lambda = \mu + 2\rho_c - \rho \quad .$$

Remarquons enfin que la restriction α_k à $\mathfrak{t}$ de toute racine Δ^+-simple α de $\Delta-\Delta_n$, est Δ_k^+-simple.

2.8. Proposition.

Les notations et hypothèses sont celles de 2.7. On suppose que $\mathfrak{B} = \mathfrak{t} \subseteq \kappa$, cest à dire que toutes les racines de Δ sont imaginaires. On définit:

$$f(\alpha) = 2\,\frac{<\lambda,\alpha>}{<\alpha,\alpha>} \quad , \ (\alpha\epsilon\Delta).$$

a/ Si α est Δ^+-simple (respmt. Δ^+-simple et compacte):

(2.8.1) $$f(\alpha) \geq -1 \quad ,(\text{respmt. } f(\alpha) \geq 1) .$$

b/ Si γ et δ sont deux racines Δ^+-simples, adjacentes et de même longueur:

(2.8.2) $$f(\gamma+\delta) \geq 0 \quad .$$

c/ Si la racine α est telle qu'il existe une racine Δ^+-simple β telle que α soit Δ_β^+-simple (voir 2.3)

(2.8.3) $$f(\alpha) \geq -1 \quad .$$

Démonstration.

On peut supposer que Δ est irréductible et différent du type G_2 , pour lequel on vérifie directement le résultat.

a/ si α est Δ^+-simple

$$f(\alpha) \geq -2\,\frac{<\rho,\alpha>}{<\alpha,\alpha>} = -1 \quad .$$

Si α est Δ^+-simple compacte, elle est Δ_k^+-simple et:

$$f(\alpha) \geq 2\frac{<2\rho_c,\alpha>}{<\alpha,\alpha>} - 2\frac{<\rho,\alpha>}{<\alpha,\alpha>} = 2 - 1 = 1 .$$

b/ Si $<\gamma,\gamma> = m<\delta,\delta>$, avec $m=1$ ou 2, on a, compte tenu des longueurs de ces diverses racines (voir [2] p.150):

$$<\gamma+\delta,\gamma+\delta> = <\delta,\delta> ,$$

(2.8.4) $$f(\gamma+\delta) = f(\delta) + mf(\gamma) ,$$

$$<\gamma+m\delta,\gamma+m\delta> = <\gamma,\gamma> ,$$

(2.8.5) $$f(\gamma+m\delta) = f(\gamma)+ f(\delta) .$$

En particulier, si $m=1$, on déduit de a/ et (2.8.4), que $f(\gamma+\delta) \geq 0$ si γ ou δ est compacte; si γ et δ sont non compactes, $\gamma+\delta$ est Δ_k^+-simple et:

(2.8.6) $$f(\gamma+\delta) \geq 2\frac{<2\rho_c,\gamma+\delta>}{<\gamma+\delta,\gamma+\delta>} - 2\frac{<\rho,\gamma+\delta>}{<\gamma+\delta,\gamma+\delta>} = 2 - 2 = 0 .$$

c/ Notons:

(2.8.7) $$\mathcal{F} = \{ \alpha\in\Delta \ / \ \exists\beta \ \Delta^+\text{-simple} , \ \alpha \text{ est } \Delta_\beta^+\text{-simple} \} .$$

Fixons $\alpha\in\mathcal{F}$ telle que:

(2.8.8) $$f(\alpha) \leq f(\alpha') , \quad (\alpha'\in\mathcal{F}) ,$$

avec

(2.8.9) $$l(\alpha) \leq l(\alpha') , \quad (\alpha'\in\mathcal{F} \text{ et } f(\alpha)=f(\alpha')).$$

Si le couple (α,β) de (2.8.7) ainsi défini vérifie les hypothèses de 2.4.c/, on a, avec les notations de 2.4 :

(2.8.10) $$w_1\alpha = \alpha - q\delta_1 , \quad (q>0),$$

et donc, d'après a/ :

$$f(w_1\alpha) = f(\alpha) - q'f(\delta_1) , \quad (q'>0) ,$$

$$f(w_1\alpha) \leq f(\alpha) ,$$

ce qui contredit le choix de α. On peut donc, d'après a/, supposer qu'on est dans le cas 2.4.b/ , avec $\beta = \gamma$ ou δ , et $\gamma = \delta$ ou γ . D'après (2.8.1-4-5) le résultat est évident si γ ou δ est compacte. si γ et δ sont non compactes, $\gamma+\delta$ est compacte et donc, comme en (2.8.6):

(2.8.11) $f(\gamma+\delta) \geq 2 - (1+2) = -1$;

(2.8.12) $f(\gamma+2\delta) = \frac{1}{2}(f(\gamma+\delta) + f(\delta)) \geq -1$.

2.9. Si β est une racine de Δ_n , on peut normaliser les vecteurs $X_{\pm\beta}$ de poids $\pm\beta$ pour que, si $H_\beta = [X_\beta, X_{-\beta}]$:

(2.9.1) $Z_\beta = X_\beta + X_{-\beta} \in \mathfrak{p}$,

(2.9.2) $Y_\beta = X_\beta - X_{-\beta} \in \sqrt{-1}\,\mathfrak{p}$,

(2.9.3) $\beta(H_\beta) = 2$.

La transformation de Cayley

(2.9.4) $\nu_\beta = \operatorname{Exp}(\frac{\pi}{4} \operatorname{ad} Y_\beta)$,

applique $\mathfrak{b}_{\mathbb{C}}$ sur une sous-algèbre de Cartan $\mathfrak{b}^1_{\mathbb{C}}$ de $\mathfrak{g}_{\mathbb{C}}$ telle que:

(2.9.5) $\mathfrak{b}^1 = \mathfrak{b}^1_{\mathbb{C}} \cap \mathfrak{g} = \operatorname{Ker} \beta + \mathbb{R} Z_\beta$;

$\mathfrak{b}^1$ est aussi une sous-algèbre de Cartan fondamentale et θ stable du centralisateur $\mathfrak{g}^1$ de Z_β dans $\mathfrak{g}$. Le système Δ^1 des racines de $(\mathfrak{g}^1,\mathfrak{b}^1)$ s'identifie, via ν_β , à Δ_β (voir 2.3). De plus, ν_β transforme une racine imaginaire α de Δ_β en une racine imaginaire $\nu_\beta(\alpha)$ de Δ^1 ; α et $\nu_\beta\alpha$ sont de même nature (compacte ou non compacte) , ou de nature différente selon que α est fortement orthogonal à β ou non (voir [7] ou [3]).

<u>2.10. Théorème.</u>(voir [10] Prop.4.1).

<u>Les notations et hypothèses sont celles de 2.7</u> .

a/ <u>Si λ n'est pas Δ^+-dominant</u> (<u>respmt. Δ^+-strictement dominant</u>) <u>il existe une racine imaginaire</u> β_1 <u>telle que si</u>

(2.10.1) $c_1 = 2\frac{<\lambda,\beta_1>}{<\beta_1,\beta_1>}$,

<u>les conditions suivantes soient réalisées:</u>

(i) β_1 <u>est Δ^+-simple ou il existe une racine Δ^+-simple et complexe</u> γ <u>telle que</u> $\beta_1 = \gamma+\theta\gamma$;

(ii) $c_1 \in]0,1]$, (respmt $c_1 \in [0,1]$) .

<u>Une telle racine</u> β_1 <u>est non compacte.</u>

b/ <u>Définissons</u>, <u>dans ce cas</u>, <u>comme en 2.9 et pour</u> $\beta = \beta_1$, $\mathfrak{g}^1$, $\mathfrak{b}^1$, Δ^1 , Δ^1_k,

(2.10.2) $\mu^1 = \mu_{|\mathfrak{g} \cap \mathfrak{b}^1}$

et $\Delta^{1,+} = \Delta^1 \cap \Delta^+$, $\Delta_k^{1,+} = \Delta_k^1 \cap \Delta^+$, ρ^1 et ρ_c^1 de manière évidente.

Alors, le système $(\mathfrak{G}^1, \mathfrak{B}^1, \mu^1, \Delta_k^{1,+}, \Delta^{1,+})$ vérifie les hypothèses de 2.7 pour

$$\lambda^1 = \mu^1 + 2\rho_c^1 - \rho^1 \quad . \tag{2.10.3}$$

c/ Soit $\{\beta_1,\ldots,\beta_p\}$ un système de racines imaginaires de Δ maximal parmi les sytèmes tels que:

(i') β_1 vérifie les conditions a/ (i)-(ii).

(ii') $\{\beta_2,\ldots,\beta_p\}$ vérifie les hypothèses de $\{\beta_1,\ldots,\beta_p\}$ pour $(\mathfrak{G}^1, \Delta^{1,+}, \lambda^1)$.

Alors:

(i")La projection $\lambda_\circ$ de λ sur le cône $\overline{\mathfrak{C}}$ des poids Δ^+-dominants est:

$$\lambda_\circ = \lambda + \frac{1}{2} \sum_{i=1}^{p} c_i \beta_i \quad , \tag{2.10.4}$$

avec:

$$c_i = -2 \frac{\langle\lambda,\beta_i\rangle}{\langle\beta_i,\beta_i\rangle} \in [0,1] \quad ,(1 \le i \le p). \tag{2.10.5}$$

(ii") L'ensemble des racines imaginaires orthogonales à $\lambda_\circ$ et au sous-espace $Sp\{\beta_1,\ldots,\beta_p\}$ engendré par $\beta_1 \ldots,\beta_p$ est vide.

Démonstration.

a/ Supposons que λ ne soit pas Δ^+-dominant; soit γ une racine Δ^+-simple telle que:

$$\langle\lambda,\gamma\rangle < 0 \quad . \tag{2.10.6}$$

D'après (2.8.1) , γ n'est pas compacte. Supposons que γ est complexe; si $\theta\gamma+\gamma$ n'est pas racine, γ et $\theta\gamma$ sont fortement orthogonales (voir (2.6.3)), et:

$$\langle\gamma_k,\gamma_k\rangle = \langle\gamma_p,\gamma_p\rangle = \frac{1}{2}\langle\gamma,\gamma\rangle \quad ,$$

de telle sorte que:

$$2\frac{\langle\lambda,\gamma\rangle}{\langle\gamma,\gamma\rangle} \ge 2\frac{\langle 2\rho_c,\gamma_k\rangle}{2\langle\gamma_k,\gamma_k\rangle} - 1 = 0 \quad .$$

$\beta_1 = \gamma+\theta\gamma$ est donc une racine imaginaire (voir 2.6) telle que $\langle\beta_1,\beta_1\rangle = \langle\gamma,\gamma\rangle = \langle\theta\gamma,\theta\gamma\rangle$, et

$$\langle\lambda,\beta_1\rangle = 2\langle\lambda,\gamma_k\rangle = 2\langle\lambda,\gamma\rangle < 0 \quad . \tag{2.10.7}$$

De plus

(2.10.8) $$2\frac{<\lambda,\beta_1>}{<\beta_1,\beta_1>} \geq 2\frac{<2\rho_c,2\gamma_k>}{2<\gamma_k,\gamma_k>} - 2\frac{<\rho,\gamma>}{<\gamma,\gamma>} - 2\frac{<\rho,\theta\gamma>}{<\theta\gamma,\theta\gamma>} = -1 .$$

Remarquons en outre que si X_{β_1} (respmt. X_γ) est un vecteur non nul de poids β_1 (respmt. γ) , on a

(2.10.11) $$X_{\beta_1} \in \mathbb{C}\,[X_\gamma,\theta X_\gamma] \subseteq \mathfrak{P}_{\mathbb{C}} ,$$

et donc β_1 est imaginaire non compacte.
Si λ est dominant, mais singulier par rapport à certaines racines imaginaires, l'ensemble

(2.10.12) $$\mathcal{F}' = \{ \beta\in\Delta^+\cap\Delta_n \;/\; <\lambda,\beta> = 0 \} ,$$

est non vide (voir 2.8.1) . Soit β un élément de $\mathcal{F}'$ rendant $l(\beta)$ minimum. Si l'assertion 2.10.a/ est fausse, il existe une racine Δ^+-simple γ telle que

(2.10.13) $$<\beta,\gamma> > 0 ,$$

de telle sorte que $\beta-\gamma \in \Delta^+$; comme λ est Δ^+-dominant et

(2.10.14) $$0 = <\beta,\lambda> = <\gamma,\lambda> + <\beta-\gamma,\lambda> ,$$

on obtient l'égalité $<\gamma,\lambda> = 0$; d'après l'hypothèse faite, ceci n'est possible que si γ est complexe; dans ce cas

$$<\beta-\gamma,\theta\gamma> \geq <\beta,\theta\gamma> = <\beta,\gamma> > 0 .$$

A nouveau, nos hypothèses se traduisent par $\beta-\gamma \neq \theta\gamma$, soit

$$\beta' = \beta - \gamma - \theta\gamma \in \mathcal{F}' ,$$

ce qui est exclu.

b/ Vérifions que μ^1 est $\Delta_k^{1,+}$-dominant. Si $\alpha\in\Delta^1$ est telle que α_k soit $\Delta_k^{1,+}$-simple, on est dans l'un des cas suivants:

- Soit α est complexe, et α_k définit une racine de Δ_k^+ , avec

$$<\mu^1,\alpha_k> = <\mu,\alpha_k> \geq 0 ;$$

-Soit α est imaginaire fortement orthogonale à β_1 ; dans ce cas, α définit une racine compacte γ de Δ^+ et

$$<\mu^1,\alpha> = <\mu,\gamma> \geq 0 ;$$

-Soit α est imaginaire non fortement orthogonale à β_1; dans ce cas, $\alpha\pm\beta_1$ appartiennent à Δ_k^+ et

$$<\mu^1,\alpha> = \frac{1}{2}\,(<\mu,\alpha+\beta_1> + <\mu,\alpha-\beta_1>) \geq 0 .$$

De même, si α est une racine $\Delta^{1,+}$-simple, elle peut être

- soit complexe et

$$<\mu^1+2\rho_c^1,\alpha> \geq <\mu^1,\alpha_k> \geq 0 \qquad ;$$

- soit imaginaire; dans ce cas, les relations (voir [5])

$$(2.10.15) \qquad 2\rho_c-\rho|_{\mathfrak{t}} = 2\rho_c^I-\rho^I|_{\mathfrak{t}} \qquad ,$$

où ρ_c^I et ρ^I sont les sommes correspondant aux racines imaginaires positives seules, et,

$$(2.10.16) \qquad 2\rho_c-\rho|_{\mathfrak{B}\cap\mathfrak{G}^1} = 2\rho_c^1-\rho^1|_{\mathfrak{B}\cap\mathfrak{G}^1}$$

permettent de se limiter au cas où toutes les racines de Δ sont imaginaires. On a alors, par application de (2.10.16) et (2.8.3) :

$$2\frac{<\mu^1+2\rho_c^1,\alpha>}{<\alpha,\alpha>} = 2\,\frac{<\lambda,\alpha>}{<\alpha,\alpha>} + 2\frac{<\rho_c^1,\alpha>}{<\alpha,\alpha>}$$

$$\geq -1 +1 = 0 \quad ,$$

de telle sorte que $\mu^1+2\rho_c^1$ est $\Delta^{+,1}$-dominant.

c/ On raisonne par récurrence sur le rang de $\mathfrak{G}$. Si λ est Δ^+-dominant, on peut supposer que toutes les racines de Δ sont imaginaires. Toutes les racines Δ^+-simples de $\mathfrak{F}'$ (voir (2.10.12)) sont non compactes et deux à deux fortement orthogonales. On conclut par application directe de 1.4 et 1.7 . Si λ n'est pas Δ^+-dominant, on écrit la décomposition définie par sa projection

$$(2.10.17) \qquad \lambda = \lambda_\circ - \sum_{\gamma\in G} c_\gamma\,\gamma \qquad ,(c_\gamma \geq 0) ,$$

où G est l'ensemble des racines Δ^+-simples orthogonales à $\lambda_\circ$. Si une composante connexe du système

$$G' = \{ \gamma\in G \;/\; c_\gamma > 0 \} \quad ,$$

de racines simples est de type A_n ,$(n\geq2)$, D_{2n+1} ,$(n\geq2)$, ou E_6 , il existe, d'après 2.5.a/ , un élément w tel que (voir (2.8.2)) :
(2.8.2)):

$$0 \geq <\lambda,w> = - \sum_{\gamma\in G} c_\gamma\,\gamma < 0 \quad .$$

On peut donc supposer (voir 2.5.b/), que si Δ' est le sous-système de Δ engendré par G', V' le sous-espace de V engendré par Δ', $\mathcal{W}'$ le sous-groupe de $\mathcal{W}$ engendré par Δ', il existe dans $\mathcal{W}'$ un élément s tel que:

(2.8.18) $$s = -Id \quad \text{sur } V'.$$

En particulier, quel que soit le système orthogonal maximal $\{\gamma_1,\dots,\gamma_q\}$ de Δ', on a (voir [6]):

(2.10.19) $$s = s_{\gamma_1}\dots s_{\gamma_q}$$

et comme $\lambda-\lambda_\circ \in V'$:

$$\lambda_\circ-\lambda = \frac{1}{2}((\lambda_\circ-\lambda)-s(\lambda_\circ-\lambda))$$
$$= \frac{1}{2}(s\lambda-\lambda)$$

(2.10.20) $$\lambda_\circ-\lambda = -\frac{1}{2}\sum_{j=1}^{q} 2\frac{<\lambda,\gamma_j>}{<\gamma_j,\gamma_j>}\gamma_j \quad .$$

Vogan construit (on supposera toutes les racines imaginaires)un système orthogonal $\{\beta_1,\dots,\beta_p\}$ de racines de Δ^+ vérifiant:

(i) β_j non compacte simple dans $\{\beta_1,\dots,\beta_{j-1}\}^\perp$, $(1\leq j\leq p)$,

(ii) $<\lambda,\beta_j> \leq 0$, $(1\leq j\leq p)$,

(iii) L'élément $\nu = \lambda - \Sigma_{j=1}^{p} \frac{<\lambda,\beta_j>}{<\beta_j,\beta_j>}\beta_j$ est dominant pour $\{\beta_1,\dots,\beta_p\}^\perp \cap \Delta^+$.

Vérifions par récurrence sur l'entier p que ν est Δ^+-dominant. Dans le cas contraire,

$$\nu = \nu_\circ - \Sigma_{\gamma\in A} c_\gamma \gamma \quad ,$$

où A est un ensemble de racines Δ^+-simples othogonales à $\nu_\circ$, $c_\gamma>0$, $(\gamma\in A)$ et $\nu_\circ$ Δ^+-dominant. D'après (1,6,1) $<\nu_\circ,\beta_1>=0$; l'hypothèse $\beta_1\notin A$ implique $<\beta_1,\gamma>=0$ pour tout $\gamma\in A$, ce qui contredit l'hypothèse de récurrence. On a donc $\beta_1\in A$ et il existe donc un système orthogonal $\{\delta_1,\dots,\delta_q\}$ de racines positives vérifiant:

(i) $\delta_1 = \beta_1$,

(ii) $<\delta_j,\nu_\circ> = 0$, $(1\leq j\leq q)$,

(iii) $$\nu = \nu_\circ + \Sigma_{j=1}^{q} \frac{<\nu,\delta_j>}{<\delta_j,\delta_j>}\delta_j \quad .$$

La relation:

$$- \Sigma_{\gamma \in A} c_\gamma \gamma = \Sigma_{j=2}^{q} \frac{<\nu, \delta_j>}{<\delta_j, \delta_j>} \delta_j \quad ,$$

implique qu'il existe un entier $j \in [2, q]$ tel que $<\nu, \delta_j> < 0$. On contredit l'hypothèse de récurrence.

2.11. Décrivons rapidement comment cette construction est utilisée par Vogan pour classifier les $(\mathfrak{g}, K)$-modules admissibles irréductibles. On applique la construction précédente à un K-type minimal μ de $\mathcal{E}$, c'est à dire un K-type figurant dans le $(\mathfrak{g}, K)$-module $\mathcal{E}$ et rendant minimal le nombre $\|\mu + 2\rho_c\|$. La transformation de Cayley (voir [5] (2.15)) associée au système orthogonal $\{\beta_1, \ldots, \beta_p\}$ définit une sous-algèbre de Cartan θ-stable $\mathfrak{h}$, telle que (voir 2.9):

(2.11.1) $$\mathfrak{h}_p = \mathfrak{h} \cap \mathfrak{p} = \mathfrak{a}_p + \sum_{j=1}^{p} \mathbb{R} Z_{\beta_j} \quad ;$$

l'espace

(2.11.2) $$\mathfrak{h}_k = \mathfrak{h} \cap \kappa = \mathfrak{t}^+$$

est l'orthogonal dans $\mathfrak{t}$ de:

(2.11.3) $$\mathfrak{t}^- = \sum_{j=1}^{p} \mathbb{R} \sqrt{-1}\, H_{\beta_j} \quad .$$

On peut choisir un parabolique cuspidal $P = MAN$ pour lequel $\mathfrak{a} = \mathfrak{h}_p$, et une représentation $(\sigma \mathcal{H}_\sigma)$ limite de série discrète de M et dont le caractère est "défini" par $\lambda_\circ$, de telle sorte que $\mathcal{E}$ soit isomorphe, pour $\nu \in \mathfrak{a}_{\mathbb{C}}^*$ convenable, à un sous-quotient de $\mathfrak{I}_{P,\sigma,\nu}$ (voir 0.2 et [1] Ch.IV) .
En particulier, le caractère infinitésimal de $\mathcal{E}$ est défini par une forme linéaire sur $\mathfrak{h}$ dont la restriction à $\mathfrak{t}^+$ est $\lambda_\circ$ (Vogan [10] Th.7.16) .

2.12 Proposition.

Les notations et hypothèses sont celles de 2.3 :

a/ $(\rho - \bar{\mathcal{C}}^\circ) \cap \bar{\mathcal{C}} = \bar{\mathcal{C}} \cap (\bigcap_{w \in W} w.(\rho - \bar{\mathcal{C}}^\circ))$.

b/ L'ensemble

$$\bigcap_{w \in W} w.(\rho - \bar{\mathcal{C}}^\circ)$$

est identique à l'enveloppe convexe Γ des points $w.\rho$, $(w \in W)$.

Démonstration.

a/ Si $\nu \in (\rho - \overline{C}^\circ) \cap \overline{C}$, on a, pour $1 \leqslant i \leqslant n$ et $w \in W$

$$\nu - w.\nu \in \overline{\mathcal{C}}^\circ$$

$$\begin{aligned} <\rho - w\nu\,, \omega_i> &= <\rho - \nu\,, \omega_i> + <\nu - w.\nu, \omega_i> \\ &\geqslant 0 \end{aligned}$$

ce qui signifie que

$$\begin{aligned} &\rho - w.\nu \in \overline{\mathcal{C}}^\circ \\ &w.\nu \in \rho - \overline{\mathcal{C}}^\circ \\ &\nu \in w^{-1}.(\rho - \overline{\mathcal{C}}^\circ) \end{aligned}$$

On a donc démontré que :

$$(\rho - \overline{\mathcal{C}}^\circ) \cap \overline{C} \subseteq \overline{\mathcal{C}} \cap \bigcap_{w \in W} w.(\rho - \overline{\mathcal{C}}^\circ)$$

L'inclusion inverse est évidente.

b/ Les ensembles concernés étant W invariants, il suffit de vérifier que

$$\Gamma \cap \overline{\mathcal{C}} = \overline{\mathcal{C}} \cap \bigcap_{w \in W} (\rho - \overline{\mathcal{C}}^\circ) = \overline{\mathcal{C}} \cap (\rho - \overline{C}^\circ)$$

Si

$$\nu = \Sigma_{w \in W}\, t_w\, w.\rho \in \Gamma \cap \overline{\mathcal{C}} \ , \quad (t_w \in [0,1]\ ,\ \Sigma_{w \in W}\, t_w = 1)\ ,$$

$$\rho - \nu = \Sigma_{w \in W}\, t_w(\rho - w.\rho) \in \overline{\mathcal{C}}^\circ$$

$$\nu \in (\rho - \overline{C}^\circ) \cap \overline{\mathcal{C}}$$

Inversement, d'après le théorème de séparation des ensembles convexes fermés, si $\nu \in \overline{C} - \Gamma$, il existe un élément μ de V tel que

$$<\mu,\nu> \; > \; <\mu, w.\rho> \quad , \quad (w \in W)$$

On peut supposer μ' dominant, car si s W est tel que :

$$\mu' = s.\mu \in \overline{\mathcal{C}}$$

on a $\mu' - s^{-1}.\mu' \in \overline{C}^\circ$ et donc :

$$\begin{aligned} <\mu',\nu> &= <\mu' - s^{-1}\mu', \nu> + <s^{-1}\mu', \nu> \\ &\geqslant <s^{-1}\mu', \nu> \\ &= <\mu, \nu> \\ &> <\mu, s^{-1}w.\rho> \\ &= <\mu', \rho> \qquad , \ (w \in W) \end{aligned}$$

Si on suppose que, de plus $\mu \in \overline{C}$

$$\begin{aligned} &<\mu,\nu> \; > \; <\mu,\rho> \\ &<\mu,\rho-\nu> \; < 0 \\ &\rho-\nu \notin \overline{\mathcal{C}}^\circ \\ &\nu \notin \rho - \overline{\mathcal{C}}^\circ \end{aligned}$$

2.13. Corollaire :

Un k-type μ ; Δ_k^+-dominant et entier, est petit si et seulement si

$$\mu + 2\rho_c \in \Gamma$$

enveloppe convexe des points $w.\rho$, $(w \in W(\mathfrak{g}_{\mathbb{C}}, \mathfrak{h}_{\mathbb{C}}))$.

Tout poids ν du k-module correspondant est poids extrémal d'un k-type petit.

Démonstration.

Le poids $\mu + 2\rho_c$ étant Δ^+ dominant, μ est petit si et seulement si

$$(\lambda_\mu)_\circ = 0$$
$$\lambda_\mu = \mu + 2\rho_c - \rho \in \overline{C}^\circ$$
$$\mu + 2\rho_c \in \Gamma \cap \overline{C}$$
$$\mu + 2\rho_c \in \Gamma \cap \overline{C}$$

Dans ce cas on peut, modulo un élément de $W(k,t)$ supposer ν Δ_k^+-dominant.

Si Δ'^+ est un système de racines positives rendant $\nu + 2\rho_c$ Δ'^+-dominant ,

$$\nu = \mu - Q \qquad \text{où } Q \in \overline{\mathcal{C}}^\circ$$

$$\begin{aligned} \lambda'_\nu &= \nu + 2\rho_c - \rho' \\ &= \mu + \rho_c - \rho'_n - Q \\ &= \rho_n - \rho'_n - \tfrac{1}{2}\Sigma_{i=1}^n c_i\alpha_i && \text{où } c_i \in [0,1] \\ &= -\Sigma_{i=1}^n c'_i\alpha'_i - Q && \text{où } \alpha'_i \in \Delta'^+ \text{ et } c'_i \in [0,1] , \end{aligned}$$

d'après un résultat de W. Schmid.

Soit $(\lambda'_\nu)_\circ = 0$

et ν est un k-type petit.

BIBLIOGRAPHIE

[1] BOREL A., WALLACH N., Continuous cohomology, discrete subgroups and representations of reductive groups, An. of Math. Studies,94,(1980), Princeton University press.

[2] BOURBAKI N., Groupes et algèbres de Lie, Ch.4-5-6,(1968), Hermann,Paris.

[3] CARMONA J., On irreducibility of the principal series, Non commutative Harmonic Analysis,Lecture Notes,587,(1977),p.1-31.

[4] HARISH CHANDRA, Harmonic Analysis on real reductive groupsIII,Ann. of Maths 104,(1976),p.117-201.

[5] HECHT H.,SCHMID W.,A proof of Blattner's conjecture, Inv. Math.,31,(1975), p.129-154.

[6] HIRAI T, The character of the discrete series for semi-simple Lie groups, Preprint, (1978).

[7] KNAPP A.W, WALLACH N.R., Szegö kernels associated with discrete series, Inv. Math.,34,(1976),p.163-200.

[8] LANGLANDS R., On the classification of irreducible representations of real algebraic groups,Preprint, Institute of Advanced Studies.

[9] MILICIC D., Asymptotic behaviour of matrix coefficients of the discrete series, Duke Math. J;,44.(1977),p.59-88.

[10] VOGAN D.A., The algebraic structure of the representations of semi-simple Lie groups I, Ann. of Maths.,109,(1979),p;1-60.

REPRESENTATIONS SPHERIQUES SINGULIERES.

par A. GUILLEMONAT

Introduction.

Dans tout ce qui suit, G désigne un groupe de Lie semi-simple de centre fini, $\underline{\underline{G}}_o = \underline{\underline{K}}_o + \underline{\underline{P}}_o$ une décomposition de Cartan de l'algèbre de Lie $\underline{\underline{G}}_o$ de G. L'étude du dual unitaire de G, même lorsqu'on se limite à la partie sphérique de ce dual, est encore loin d'être achevée. La détermination des conditions nécessaires et suffisantes d'irréductibilité des séries dégénérées est aussi un problème non résolu. Les résultats les plus importants sur le dual unitaire de classe un, sont dûs à B. Kostant, (voir [5]) qui donne un critère nécessaire et suffisant d'irréductibilité des représentations induites par une représentation sphérique de dimension un d'un parabolique minimal.
Il détermine aussi une zone critique unitaire, qui n'est qu'une partie seulement du dual cherché. Une généralisation assez large du problème est de chercher des critères nécessaires et suffisants d'irréductibilité de l'induite à G d'une représentation de dimension un d'un parabolique quelconque B de G d'algèbre $\underline{\underline{B}}_o$. Nous donnons ici une réponse compléte à ce problème, pour une large classe de groupes (dont par exemple, $SL(n,\mathbb{C})$, $SL(n,\mathbb{R})$, $SU^*(2n)$, $n \in \mathbb{N}$, etc...). En fait, nous montrerons que le problème général se ramène au cas où B est maximal. Nous étudions ici essentiellement le cas où la paire $(\underline{\underline{K}}_o, \underline{\underline{B}}_o \cap \underline{\underline{K}}_o)$ est symétrique, et nos méthodes nous permettent, pour ce type de représentations, de donner également un critère nécessaire et suffisant d'unitarisation. Nous utilisons largement dans ce qui suit nos articles précédents (voir [1] et [2]) où un problème semblable avait été étudié pour une paire hermitienne symétrique de type tubulaire.
Les notations sont cumulatives.

1. Irréductibilité des modules sphériques induits.

1.1 - Etant donnée une algèbre de Lie simple réelle $\underline{\underline{G}}_o$, on fixe une conjugaison de Cartan θ de $\underline{\underline{G}}_o$ et on note

(1.1.1) $$\underline{\underline{G}}_o = \underline{\underline{K}}_o + \underline{\underline{P}}_o$$

la décomposition de Cartan correspondante de $\underline{\underline{G}}_o$, $\underline{\underline{K}}_o$ étant l'ensemble des points fixes de θ . On note B la forme de Killing de $\underline{\underline{G}}_o$, $\underline{\underline{G}}$ la complexifiée de $\underline{\underline{G}}_o$, $\underline{\underline{E}}$ le complexifié de tout sous espace $\underline{\underline{E}}_o$ de $\underline{\underline{G}}_o$, $X \to \overline{X}$ la conjugaison de $\underline{\underline{G}}$ définie par $\underline{\underline{G}}_o$. On prolonge B et θ par $\mathbb{C}$-linéarité. Pour toute algèbre de Lie $\mathcal{L}$ (respect. tout groupe de Lie L) et tout $\underline{\underline{G}}$-module (respect. L-module) V , on note

(1.1.2) $$V^{\mathcal{L}} = \{ v \in V \ / \ \forall\, X \in \mathcal{L} \qquad X.v = 0 \}$$

(1.1.3) $$V^{L} = \{ v \in V \ / \ \forall\, x \in L \qquad x.v = v \} \ .$$

Dans tout ce qui suit, orthogonal signifie orthogonal relativement à B .

1.2 - Lorsqu'une algèbre de Lie $\mathcal{L}$ (respect. un groupe de Lie L) agit sur un espace vectoriel de dimension finie V , on étend l'action de $\mathcal{L}$ (respect. L) à l'algèbre symétrique S(V) de V par des dérivations (respect. des automorphismes) d'algèbre de S(V) . L'action obtenue laisse invariant le sous-espace $S(V)_m$ des éléments homogènes de degré m de S(V) .

En particulier, si K est le sous-groupe du groupe adjoint de $\underline{\underline{G}}_o$ correspondant à l'algèbre de Lie ad $\underline{\underline{K}}_o$, on fera agir K et $\underline{\underline{K}}$ sur $S(\underline{\underline{P}})$.

En outre, on étend à cette algèbre la forme bilinéaire symétrique sur $\underline{\underline{P}}$ définie par la forme de Killing B de $\underline{\underline{G}}$.

On note $F : S(\underline{\underline{G}}) \to \underline{\underline{U}}(\underline{\underline{G}})$ l'application canonique.

1.3 - Dans tout ce qui suit, on fixe une sous-algèbre parabolique

(1.3.1) $$\underline{\underline{B}}_o = \underline{\underline{M}}_o + \underline{\underline{A}}_o + \underline{\underline{N}}_o \ , \quad \text{(décomposition de Langlands)} \ ,$$

de radical nilpotent $\underline{\underline{N}}_o$. L'algèbre réductive $\underline{\underline{M}}_o$ centralise le sous-espace abélien $\underline{\underline{A}}_o$ de $\underline{\underline{P}}_o$; elle est invariante par θ de telle sorte que

(1.3.2) $$\underline{\underline{M}}_o = \underline{\underline{K}}'_o + \underline{\underline{P}}'_o \ ,$$

où $\underline{\underline{K}}'_o = \underline{\underline{K}}_o \cap \underline{\underline{M}}_o$ et $\underline{\underline{P}}'_o = \underline{\underline{P}}_o \cap \underline{\underline{M}}_o$, est une décomposition de Cartan de $\underline{\underline{M}}_o$.

On note K' le sous-groupe de K (voir 1.2) d'algèbre $\mathrm{ad}_{\underline{\underline{G}}}\, K'_o$, M le centralisateur de $\underline{\underline{A}}$ dans K et ρ l'élément du dual $\underline{\underline{A}}'$ de $\underline{\underline{A}}$ tel que

$$(1.3.3) \qquad \rho(H) = \frac{1}{2}\,\mathrm{Tr}\,\mathrm{ad}\,H|_{N} \quad , \ (H \in \underline{\underline{A}}) \ .$$

1.4 - On fixe une sous-algèbre parabilique minimale

$$(1.4.1) \qquad {}^{\circ}\underline{\underline{B}}_{\circ} = {}^{\circ}\underline{\underline{M}}_{\circ} + {}^{\circ}\underline{\underline{A}}_{\circ} + {}^{\circ}\underline{\underline{N}}_{\circ} \quad ,$$

contenue dans $\underline{\underline{B}}_{\circ}$, avec $\underline{\underline{A}}_{\circ} \subset {}^{\circ}\underline{\underline{A}}_{\circ}$, ${}^{\circ}\underline{\underline{M}}_{\circ} \subset \underline{\underline{M}}_{\circ}$ et $\underline{\underline{N}}_{\circ} \subset {}^{\circ}\underline{\underline{N}}_{\circ}$, et on définit ${}^{\circ}\rho \in {}^{\circ}\underline{\underline{A}}'$ de façon analogue. Soient $\tau : S(\underline{\underline{A}}) \to S(\underline{\underline{A}})$ (respect. ${}^{\circ}\tau : S({}^{\circ}\underline{\underline{A}}) \to S({}^{\circ}\underline{\underline{A}})$) , l'automorphisme d'algèbre unitaire tel que

$$(1.4.2) \qquad \tau(H) = H - \rho(H)\,1 \quad , \ (H \in \underline{\underline{A}}) \ ,$$

$$(1.4.3) \qquad (\text{respect. } {}^{\circ}\tau(H) = H - {}^{\circ}\rho(H)\,1 \quad , \ (H \in {}^{\circ}A)\ )) \ ,$$

et $\zeta : \underline{\underline{U}}(\underline{\underline{G}}) \to S(\underline{\underline{A}})$, $\xi : \underline{\underline{U}}(\underline{\underline{G}}) \to S({}^{\circ}\underline{\underline{A}})$) les applications définies par

$$(1.4.4) \qquad u = \zeta(u) \ \mathrm{mod}\ (\underline{\underline{K}}\,\underline{\underline{U}}(\underline{\underline{G}}) + \underline{\underline{U}}(\underline{\underline{G}})\,(\underline{\underline{M}} + \underline{\underline{N}})) \ , \ (u \in \underline{\underline{U}}(G)) \ ,$$

$$(1.4.5) \qquad u = \xi(u) \ \mathrm{mod}\ (\underline{\underline{K}}\,\underline{\underline{U}}(\underline{\underline{G}}) + \underline{\underline{U}}(\underline{\underline{G}})\,{}^{\circ}\underline{\underline{N}}) \ , \ (u \in \underline{\underline{U}}(\underline{\underline{G}})) \ ,$$

où

$$(1.4.6) \qquad \underline{\underline{A}}_{\circ}^{\perp} = {}^{\circ}\underline{\underline{A}} \cap \underline{\underline{M}}_{\circ} \quad ,$$

est abélien maximal dans $\underline{\underline{P}}_{\circ}'$. On sait que :

$$(1.4.7) \qquad \xi(uv) = \xi(u)\,\xi(v) \quad , \ (u \in \underline{\underline{U}}(\underline{\underline{G}})^{K} , \ v \in \underline{\underline{U}}(\underline{\underline{G}})) \ .$$

On identifiera le groupe de Weyl W de $(\underline{\underline{M}},\underline{\underline{A}}^{\perp})$ à un sous-groupe du groupe de Weyl ${}^{\circ}W$ de $(\underline{\underline{G}},{}^{\circ}\underline{\underline{A}})$ formé d'éléments qui agissent trivialement sur $\underline{\underline{A}}$. On sait (voir par exemple Kostant [5]) que

$$(1.4.8) \qquad \xi' = {}^{\circ}\tau_{\circ}\,\xi$$

définit un homomorphisme de $\underline{\underline{U}}(\underline{\underline{G}})^{K}$ sur $S({}^{\circ}\underline{\underline{A}})^{{}^{\circ}W}$ de noyau

$$(1.4.9.) \qquad \underline{\underline{U}}(\underline{\underline{G}})^{K} \cap \underline{\underline{K}}\,\underline{\underline{U}}(\underline{\underline{G}}) = \underline{\underline{U}}(\underline{\underline{G}})^{K} \cap \underline{\underline{U}}(\underline{\underline{G}})\,\underline{\underline{K}} \ .$$

1.5 - A toute forme linéaire $\mu : {}^{\circ}\underline{\underline{A}} \to \mathbb{C}$, Kostant associe l'idéal à gauche

$$(1.5.1) \qquad J_{\mu} = \underline{\underline{U}}(\underline{\underline{G}})\,\underline{\underline{K}} + \Sigma_{u \in \underline{\underline{U}}(\underline{\underline{G}})^{K}}\ \underline{\underline{U}}(\underline{\underline{G}})\,(u - \xi'(u)(\mu)\,1) \quad ,$$

et le $\underline{\underline{G}}$ - module sphérique

$$(1.5.2) \qquad Y_\mu = \underline{\underline{U}}(\underline{\underline{G}})/_{J_\mu}$$

L'idéal J_μ est contenu dans un idéal à gauche maximal unique J'_μ de telle sorte que le $\underline{\underline{G}}$ - module sphérique

$$(1.5.3) \qquad Y'_\mu = \underline{\underline{U}}(\underline{\underline{G}})/_{J'_\mu} \quad ,$$

est l'unique quotient simple de Y_μ . D'après (1.4) :

$$(1.5.4) \qquad J_{w\mu} = J_\mu \qquad , (w \in {}^\circ W) \quad ,$$

et donc :

$$(1.5.5) \qquad J'_{w\mu} = J'_\mu \ , \ Y_\mu = Y_{w\mu} \ , \ Y'_\mu = Y'_{w\mu} \ .$$

Pour u dans ${}^\circ\underline{\underline{A}}'$, il existe une application sesquilinéaire unique $(v,w) \to \langle v,w \rangle$ de $Y_\mu \times Y_{-\bar{\mu}}$ dans $\mathbb{C}$, $\underline{\underline{G}}$ - invariante et telle que

$$(1.5.6) \qquad \langle \dot{1} , \ddot{1} \rangle = 1$$

où $\dot{u}$ (respect. $\ddot{u}$) est la classe de $u \in U(G)$ dans Y_μ (respect. $Y_{-\bar{\mu}}$).
Plus précisément, si dk est la mesure de Haar normée de K

$$(1.5.7) \qquad \langle \dot{u} , \ddot{v} \rangle = \xi'(\int_K (k.v)^* (k.u)\, dk)(\mu) \quad , (u,v \in \underline{\underline{U}}(\underline{\underline{G}})) \ ,$$

où $u \to u^*$ est l'antiautomorphisme semi-linéaire de $\underline{\underline{U}}(\underline{\underline{G}})$ tel que :

$$(1.5.8) \qquad X^* = - \overline{X} \qquad , (X \in \underline{\underline{G}}(\underline{\underline{G}})) \ .$$

En particulier, il existe sur Y_μ une forme sesquilinéaire $\underline{\underline{G}}$ - invariante si et seulement si :

$$(1.5.9) \qquad \exists\, w \in {}^\circ W \qquad w.\mu = - \bar{\mu}$$

<u>1.6 Lemme</u> :

a) Pour tout élément u de $\underline{\underline{U}}(\underline{\underline{G}})^{K'}$

$$(1.6.1) \qquad \xi'(u) \in S({}^\circ\underline{\underline{A}})^W$$

b) Plus généralement, pour tout élément u de $\underline{\underline{U}}(\underline{\underline{G}})$

$$(1.6.2) \qquad \xi(u) = \xi(\int_{K'} k'.u\, dk') \mod S({}^\circ\underline{\underline{A}})\, \underline{\underline{A}}^\perp$$

où dk' est la mesure de Haar normée de K' .

Démonstration.

a) Soit, pour $u \in \underline{\underline{U}}(\underline{\underline{G}})$, u' l'unique élément de $U(\underline{\underline{A}})\ F(S(\underline{\underline{P}}'))$, (voir (1.2)), tel que

(1.6.3) $u = u' \mod (\underline{\underline{K}}\,\underline{\underline{U}}(\underline{\underline{G}}) + \underline{\underline{U}}(\underline{\underline{G}})\,\underline{\underline{N}})$

Cette décomposition étant unique et K' invariante

(1.6.4) $k'.u' = u'$, $(k' \in K')$

et,

(1.6.5) $\xi(u') = \xi(u)$

Il suffit donc d'appliquer à $\underline{\underline{M}}$ le résultat de 1.4 en remarquant que

(1.6.6) $\frac{1}{2}\,\mathrm{Tr}\ \mathrm{ad}\ H|_{{}^\circ\underline{\underline{N}} \cap \underline{\underline{M}}} = {}^\circ\rho(H)$, $(H \in \underline{\underline{A}}^{\perp})$

b) De même, si u et u' sont définis par 1.6.3 , on a pour $k' \in K'$

(1.6.7) $k'.u = k'.u' \mod (\underline{\underline{K}}\,\underline{\underline{U}}(\underline{\underline{G}}) + \underline{\underline{U}}(\underline{\underline{G}})\,\underline{\underline{N}})$

D'autre part, si $v \in S(P')$ est homogène de degré strictement positif

(1.6.8) $\xi(F(v)) \in S({}^\circ\underline{\underline{A}})\ \underline{\underline{A}}^{\perp}$,

Il en résulte que

(1.6.9) $\zeta(k'.u) = \zeta(u')$

On a donc

(1.6.10) $\zeta(\int_{K'} k'.u\ dk') = \zeta(u')$

(1.6.11) $\xi(\int_{K'} k'.u\ dk') = \xi(u) \mod S({}^\circ\underline{\underline{A}})\ \underline{\underline{A}}^{\perp}$

1.7 Remarque :

Si $u \in \underline{\underline{U}}(\underline{\underline{G}})$ appartient à un sous-espace K-stable V tel que $V^{K'} = \{0\}$, on a

(1.7.1) $\int_{K'} k'.u\ dk' = 0$

et donc

(1.7.2) $\xi(u) \in S({}^\circ\underline{\underline{A}})\ A^{\perp}$

1.8 Soit ${}^\circ M$ le centralisateur de ${}^\circ\underline{\underline{A}}$ dans K . Pour tout K-module irréductible V , la multiplicité de V dans Y_μ est égale à la dimension n de $V^{{}^\circ M}$.

1.9 On identifie toute forme linéaire sur $\underline{\underline{A}}_\circ$ (resp $\underline{\underline{A}}_\circ^\perp$) à une forme linéaire sur $^\circ\underline{\underline{A}}$ nulle sur $\underline{\underline{A}}_\circ^\perp$ (resp $\underline{\underline{A}}_\circ$).

En particulier

$$^\circ\rho = \rho + \rho' \quad ,$$

où $\rho' = {}^\circ\rho|_{\underline{\underline{A}}^\perp}$. On fixe désormais une forme linéaire ν sur $\underline{\underline{A}}_\circ$, à valeurs dans $\mathbb{C}$ et on supposera par la suite que :

$$(1.9.1) \qquad \mu = \rho' + \nu \quad .$$

1.10 Soit $\widetilde{G}$ un groupe de Lie connexe de centre fini d'algèbre de Lie $\underline{\underline{G}}_\circ$, $\widetilde{B}$ le normalisateur de $\underline{\underline{B}}_\circ$ dans G. On écrit $\widetilde{B} = \widetilde{M}\widetilde{A}\widetilde{N}$ la décomposition de Langlands de $\widetilde{B}$, où $\widetilde{M}\widetilde{A} = \widetilde{B} \cap \theta\widetilde{B}$.

Soit $\underline{\underline{H}}_\mu$ l'espace de la représentation induite $\mathrm{Ind}_{\widetilde{B}\uparrow\widetilde{G}}\widetilde{T}_\nu$, c'est-à-dire l'espace des fonctions $f : G \to \mathbb{C}$, $\underline{\underline{K}}$ finies à gauche et vérifiant :

$$(1.10.1) \qquad f(gman) = a^{\mu-\rho} f(g) \;, \; (g \in \widetilde{G}, m \in \widetilde{M}, a \in \widetilde{A}, n \in \widetilde{N})$$

On définit de façon standard une action de $\underline{\underline{G}}$, à gauche dans H_μ qui devient ainsi un $\underline{\underline{G}}$-module sphérique.

1.11 <u>Théorème</u>.

Si $\underline{\underline{H}}_\mu$ est irréductible, il est isomorphe à Y'_μ.

<u>Preuve</u> Si v est un vecteur sphérique non nul de $\underline{\underline{H}}_\mu$, on a

$$(1.11.1) \qquad J_\mu . v = 0$$

L'application

$$(1.11.2) \qquad a \to a.v$$

de $U(\underline{\underline{G}})$ dans $\underline{\underline{H}}_\mu$ passe donc au quotient en un $(\underline{\underline{G}},K)$-morphisme non nul :

$$(1.11.3) \qquad Y_\mu \to H_\mu \;.$$

D'autre part, si $\underline{\underline{H}}_\mu$ est irréductible, l'image de J'_μ est nulle car le $\underline{\underline{G}}$-module J'_μ / J_μ n'est pas sphérique.

1.12 <u>Remarque</u> Pour tout $\widetilde{K}$-type inductible V la multiplicité $m(V)$ de

V dans $\underline{\underline{H}}_\mu$, vérifie :

$$(1.12.1) \qquad m(V) = \dim V^{\widetilde{M}} \cap \widetilde{Y}_1$$

En particulier, si $m'(V)$ est la multiplicité de V dans Y'_μ , $\underline{\underline{H}}_\mu$ sera inductible si et seulement si, quel que soit le K-type V :

$$(1.12.2) \qquad m'(V) = \dim V^{\widetilde{M}_\circ \widetilde{K}} ,$$

En effet, d'après (1.12.1) la condition est nécessaire. Elle est suffisante car d'après la démonstration de (1.8), Y'_μ est infinitésimalement équivalent à un sous-quotient de $\underline{\underline{H}}_\mu$, à savoir le quotient de l'image de Y_μ par l'image de J'_μ .

1.13 - Soit $\underline{\underline{B}}'_\circ$ une sous-algèbre parabolique de $\underline{\underline{G}}_\circ$ admettant une décomposition de Langlands à facteur de Levi θ-stable,

$$(1.13.1) \qquad \underline{\underline{B}}'_\circ = \underline{\underline{M}}''_\circ + \underline{\underline{A}}'_\circ + \underline{\underline{N}}'_\circ ,$$

où $\qquad \underline{\underline{M}}''_\circ + \underline{\underline{A}}'_\circ = \underline{\underline{B}}'_\circ \cap \theta B'_\circ$, avec :

$$(1.13.2) \qquad \underline{\underline{M}}''_\circ \supset \underline{\underline{M}}_\circ \ ; \ \underline{\underline{M}}''_\circ + \underline{\underline{A}}'_\circ \supset \underline{\underline{M}}_\circ + \underline{\underline{A}}_\circ ,$$

et

$$(1.13.3) \qquad \dim \underline{\underline{A}}'_\circ = \dim \underline{\underline{A}}_\circ - 1$$

Si

$$(1.13.4) \qquad \underline{\underline{M}}'_\circ = [\underline{\underline{M}}''_\circ , \underline{\underline{M}}''_\circ] ,$$

la sous-algèbre parabolique maximale $\underline{\underline{B}}_1 = \underline{\underline{M}}'_\circ \cap \underline{\underline{B}}_\circ$ de $\underline{\underline{M}}'_\circ$ admet une décomposition de Langlands :

$$(1.13.5) \qquad \underline{\underline{B}}_1 = \underline{\underline{M}}_\circ + \underline{\underline{A}}_1 + \underline{\underline{N}}_1$$

avec

$$\underline{\underline{A}}_1 = \underline{\underline{A}}_\circ \cap \underline{\underline{M}}'_\circ \subset \underline{\underline{A}}_\circ , \quad \underline{\underline{N}}_1 = \underline{\underline{N}}_\circ \cap \underline{\underline{M}}'_\circ$$

de telle sorte que (voir (1.13.3)) :

$$(1.13.6) \qquad \dim \underline{\underline{A}}_1 = 1 .$$

Soient M' le sous-groupe connexe de $\widetilde{G}$ d'algèbre de Lie $\underline{\underline{M}}'_\circ$, $\widetilde{B}_1$ le normalisateur de $\underline{\underline{B}}_1$ dans $\widetilde{M}'$, et :

$$\rho_1(A) = \frac{1}{2} \operatorname{Tr} \operatorname{ad} A_{|\underline{\underline{N}}_1} \quad , \quad (A \in \underline{\underline{A}}_1) .$$

Nous démontrerons dans un prochain article le résultat suivant.

1.14 - Théorème

La représentation induite de $\widetilde{B} = \widetilde{M}\widetilde{A}\widetilde{N}$ à $\widetilde{G}$: $\operatorname{Ind}_{\widetilde{B}\uparrow\widetilde{G}} T_\mu$ est réductible si et seulement si il existe une sous-algèbre parabolique $\underline{\underline{B}}'_\circ$, vérifiant (1.13.2) et (1.13.3) , telle que la représentation induite de $\widetilde{B}_1$ à $\widetilde{M}'$,$\operatorname{Ind}_{\widetilde{B}\uparrow\widetilde{M}'} T^1_\mu$ est réductible , avec

$$T^1_\mu(b_1) = a_1^{\mu-\rho_1}$$

(où $b_1 = m_1 a_1 n_1$ est la décomposition de Langlands de $b_1 \in \widetilde{B}_1$) .

1.15 - Le théorème 1.14 met en évidence l'intérêt tout particulier de l'étude des représentations induites par une représentation de dimension un d'un parabolique maximal $B_\circ$. Nous allons faire une étude compléte de ce problème en nous restreignant dans cet article au cas où la paire $(\underline{\underline{K}}_\circ , \underline{\underline{K}}'_\circ)$ est symétrique. Supposons la sous-algèbre parabolique $\underline{\underline{B}}_\circ$ (voir 1.3) maximale dans $\underline{\underline{G}}_\circ$. Soit $\psi_\circ$ la racine restreinte $^\circ\underline{\underline{B}}$ -simple non nulle sur $A_\circ$,$\psi_1,..\psi_{\ell-1}$, $(\ell = \dim {}^\circ\underline{\underline{A}})$ les autres racines restreintes $^\circ\underline{\underline{B}}$-simples.

1.16 - Proposition

La paire $(\underline{\underline{K}}_\circ,\underline{\underline{K}}'_\circ)$ est symétrique si et seulement si toute racine restreinte positive φ s'écrit :

$$(1.16.1) \qquad \varphi = \sum_{i=1}^{\ell-1} n_i \psi_i + n_\circ\psi_\circ$$

avec

$$(1.16.2) \qquad n_\circ = 0 \text{ ou } 1 .$$

Preuve : Soit $\underline{\underline{U}}_1$ le commutant de $\underline{\underline{A}}_\circ$ dans $\underline{\underline{G}}_\circ$.

Soit pour une racine positive (c'est-à-dire positive relativement à $°\underline{\underline{B}}_\circ$) λ de $(\underline{\underline{G}}_\circ,\underline{\underline{A}}_\circ)$ non nulle sur $\underline{\underline{A}}_\circ$, $\underline{\underline{P}}^+_\lambda$ (resp $\underline{\underline{P}}^-_\lambda$) l'espace vectoriel réel sur $\mathbb{R}$ formé des vecteurs X de $\underline{\underline{N}}_\circ$ de poids λ sur $\underline{\underline{A}}_\circ$ (resp $-\lambda$) .
Soit $\lambda_\circ$ le plus haut poids sur $\underline{\underline{A}}_\circ$ de $\underline{\underline{N}}_\circ$. L'espace $\underline{\underline{G}}'_\circ$ défini par :

$$(1.14.2)\qquad \underline{\underline{G}}'_\circ = \underline{\underline{U}}_1 + (\underline{\underline{P}}^+_{\lambda_\circ} + P^-_{\lambda_\circ})$$

est une algèbre de Lie réductive, la paire $(\underline{\underline{G}}'_\circ, \underline{\underline{U}}_1)$ est symétrique et (1.14.2) définit une décomposition de Cartan de $\underline{\underline{G}}'_\circ$ associée à une involutive σ commutant avec θ . La trace sur $\underline{\underline{K}}_\circ$ de $\underline{\underline{G}}'_\circ$ définit donc une décomposition de Cartan de $\underline{\underline{G}}'_\circ \cap \underline{\underline{K}}_\circ$ dont l'ensemble des points fixes est $\underline{\underline{K}}'_\circ$ puisque en fait $\underline{\underline{U}}_1 = \underline{\underline{M}}_\circ + \underline{\underline{A}}_\circ$. Clairement, la condition (1.14.1) signifie que $\underline{\underline{G}}'_\circ = \underline{\underline{G}}_\circ$ ce qui équivaut à $\underline{\underline{G}}'_\circ \cap \underline{\underline{K}}_\circ = \underline{\underline{K}}_\circ$. La réciproque résulte du fait que si la paire $(\underline{\underline{K}}_\circ,\underline{\underline{M}}_\circ \cap \underline{\underline{K}}_\circ)$ est symétrique $\underline{\underline{M}}_\circ \cap \underline{\underline{K}}_\circ$ est une sous-algèbre réductive maximale de $\underline{\underline{K}}_\circ$ distincte de $\underline{\underline{G}}'_\circ \cap \underline{\underline{K}}_\circ$.

Q.E.D.

1.17 - Corollaire

Soient $\underline{\underline{G}}_\circ,\underline{\underline{B}}_\circ,\underline{\underline{B}}'_\circ$ vérifiant les hypothèses de 1.13 . On supposera que $\underline{\underline{G}}_\circ$ admet un système de racines restreintes de type a_n . Alors avec les notations de 1.13 la paire $(\underline{\underline{M}}_\circ \cap \underline{\underline{K}}_\circ , \underline{\underline{M}}''_\circ \cap \underline{\underline{K}}_\circ)$ est symétrique.

Preuve : Pour toute racine restreinte simple $\psi_\circ$ la condition (1.15.1) est vérifiée. Bien entendu, un tel système est réduit.

Q.E.D.

1.18 - Remarque. En vertu du Corollaire 1.17 et du Théorème 1.14 , le problème des singularités des représentations induites par la représentation T_μ d'un parabolique quelconque $\underline{\underline{B}}_\circ$ sera complètement résolue par nos résultats pour les algèbres sur $\mathbb{R}$ admettant un système de racines restreintes de type a_n . Citons par exemple les types : AI : $(s\ell(n,\mathbb{R})$, $so(n))$,

AII ($sn^*(2n)$, $sp(n)$) , $s\ell(n,\mathbb{C})$.

1.18. Supposons dim $\underline{\underline{A}}_o = 1$, cherchons à quelle condition μ vérifiant (1.9.1) vérifie (1.5.9) .

Nous allons expliciter cette condition dans le cas étudié actuellement, en particulier :

$$(1.18.1) \qquad \mu = \nu + \rho'$$

ν pouvant s'identifier à une forme linéaire sur ${}^o\underline{\underline{A}}_o$ nulle sur $\underline{\underline{A}}_o^{\perp}$

Soit H_o un élément non nul de $\underline{\underline{A}}_o$.

1.19 Proposition :

Pour que le module sphérique irréductible associé à la forme μ définie en (1.12) soit hermitien ,il est nécessaire que ν soit réel ou imaginaire pur. Si ν est réel non nul, on considère deux cas :

(i) S'il existe un élément w_1 de ${}^oW$ tel que :

$$(1.19.1) \qquad w_1(H_o) = - H_o \quad , 0 \neq H_o \in A_o ,$$

tout ν réel définit un module sphérique hermitien.

(ii) Dans le cas contraire, il faut et il suffit que μ annule tous les générateurs homogènes de degré impair de l'algèbre $S({}^o\underline{\underline{A}})^{{}^oW}$ des ${}^oW$-invariants. Dans ce cas, il n'y a qu'un nombre fini de points hermitiens réels.

Démonstration.

La condition (1.5.9) s'écrit :

$$w\,\mathrm{Im}\,\nu = \mathrm{Im}\,\nu ,$$

$$w(\mathrm{Re}\,\nu+\rho') = -(\mathrm{Re}\,\nu+\rho') ,$$

Si $\mathrm{Im}\,\nu \neq 0$, w induit l'identité sur $\mathbb{R}H_o$ et donc :

$$w\,\mathrm{Re}\,\nu = -\,\mathrm{Re}\,\nu$$

$$w\rho' = -\rho'$$

On a donc $\mathrm{Re}\,\nu = 0$. Cette condition est suffisante, car il existe un élément w_o de W tel que :

$$(1.19.2) \qquad w_o(H_o) = H_o ,$$

$$(1.19.3) \qquad w_o(\rho') = -\rho' .$$

Supposons ν réel non nul. Dans le cas (i) fixons $w_2 \in W$ tel que :

$$w_2 \nu = \nu \ ,$$

$$w_2 w_1 \rho' = -\rho' \ ,$$

l'élément $w_2 w_1 \in W$ vérifie (1.5.1) . Dans le cas (ii) , la condition proposée est une condition nécessaire et suffisante pour que :

$$\forall\ a \in S(^{\circ}\underline{\underline{A}}_{\circ})^{^{\circ}W} \qquad a(\nu+\rho') = a(-(\nu+\rho')) \ .$$

On sait que cette condition est équivalente à (1.5.9) . Enfin, s'il existe une infinité de points hermitiens réels, il existe un élément $w \in {}^{\circ}W$ et deux formes distinctes ν_1 et ν_2 telles que :

$$w(\nu_1+\rho') = -(\nu_1+\rho') \ ,$$

$$w(\nu_2+\rho') = -(\nu_2+\rho') \ ,$$

et donc

$$w(\nu_1-\nu_2) = -(\nu_1-\nu_2) \ .$$

On est donc dans le cas (i) .

<u>Remarque</u> : Les points $\mu = \rho'$, $\mu = \pm\rho+\rho'$ sont toujours hermitiens. La situation (ii) est effectivement rencontrée dans le cas $G = SO(2n,\mathbb{C})$ avec n impair.

1.21 - <u>Remarque</u> Si μ vérifie (1.5.9) , il est clair que Y'_{μ} est infinitésimalement unitaire ssi pour tout $v \in \underline{\underline{U}}(\underline{\underline{G}})$ l'algèbre enveloppante de $\underline{\underline{G}}$.

$$(1.21.1) \qquad \xi'(\int_K (kv) * (kv)\, dk)(\nu+\rho') \geqslant 0 \ .$$

Nous expliciterons complètement cette condition dans le cas où $\dim \underline{\underline{A}}_{\circ} = 1$, $(\underline{\underline{K}}_{\circ},\underline{\underline{K}}'_{\circ})$ étant une paire symétrique.

2. Quelques propriétés du $\underline{\underline{K}}$-module $\underline{\underline{G}}$.

2.1 - On fixe une sous-algèbre de Cartan $\underline{\underline{T}}_\circ$ de $\underline{\underline{K}}_\circ$ et on note Φ (respect. Φ_k , respect. Φ_n) l'ensemble des poids de $\underline{\underline{T}}$ dans $\underline{\underline{G}}$ (respect. $\underline{\underline{K}}$, respect. $\underline{\underline{P}}$) . Pour tout $\nu \in \Phi_k$ (respect. $\nu \in \Phi_n$) non nul, le sous-espace de poids $\underline{\underline{P}}^\nu$ de $\underline{\underline{P}}$ (respect. $\underline{\underline{K}}^\nu$ de $\underline{\underline{K}}$) et de dimension un et

(2.1.1) $$[\underline{\underline{K}}^\nu , \underline{\underline{K}}^{-\nu}] = [\underline{\underline{P}}^\nu , \underline{\underline{P}}^{-\nu}] = \mathbb{C}\, \underline{\underline{H}}_\nu$$

où H_ν est le vecteur de $\underline{\underline{T}}_{\mathbb{R}} = \sqrt{-1}\, \underline{\underline{T}}_\circ$, orthogonal au noyau de ν et tel que :

(2.1.2) $$\nu(H_\nu) = 2$$

Enfin, pour $\nu \in \Phi_n$ (respect. $\nu \in \Phi_k$) on choisira un vecteur non nul $X_\nu \in \underline{\underline{P}}^\nu$ (respect. $Y_\nu \in \underline{\underline{K}}^\nu$) de telle sorte que $\{X_\nu, X_{-\nu}, H_\nu\}$ (respect. $\{Y_\nu, Y_{-\nu}, H_\nu\}$) soit une base canonique d'une T.D.S. (voir Kostant [5]) . Remarquons enfin que, pour tout poids non nul ν de Φ

(2.1.3) $$2\nu \in \Phi \Rightarrow \nu \in \Phi_n \cap \Phi_k \quad \text{et} \quad 2\nu \in \Phi_n - \Phi_k$$

2.2 - Soit σ un automorphisme involutif de $\underline{\underline{K}}_\circ$, $\underline{\underline{M}}_\circ$ l'ensemble des points fixes de σ et

(2.2.1) $$\underline{\underline{K}}_\circ = \underline{\underline{M}}_\circ + \underline{\underline{Q}}_\circ \quad ,$$

la décomposition de Cartan correspondante. Dans tout ce qui suit, on choisit $\underline{\underline{T}}_\circ$ invariante par σ et

(2.2.2) $$\underline{\underline{T}}_\circ^- = \underline{\underline{T}}_\circ \cap \underline{\underline{Q}}_\circ \quad ,$$

abélienne maximale dans $\underline{\underline{Q}}_\circ$, de telle sorte que, si $\underline{\underline{T}}_\circ^+ = \underline{\underline{T}}_\circ \cap \underline{\underline{M}}_\circ$:

(2.2.3) $$\underline{\underline{T}}_\circ = \underline{\underline{T}}_\circ^+ + \underline{\underline{T}}_\circ^- \quad ,$$

est une somme directe orthogonale.

2.3 - On dira qu'un ordre sur le dual réel de $\underline{\underline{T}}_{\mathbb{R}}$ est compatible si cet ordre est l'ordre lexicographique associé à une base $\{H_1,\ldots,H_t\}$ de $\underline{\underline{T}}_{\mathbb{R}}$ où $\{H_1,\ldots,H_s\}$ est une base de $\underline{\underline{T}}_{\mathbb{R}}^- = \underline{\underline{T}}_{\mathbb{R}} \cap \underline{\underline{T}}^-$ et $\{H_{s+1},\ldots,H_t\}$ une base de

$\underline{\underline{T}}^{+}_{\mathbb{R}} = \underline{\underline{T}}_{\mathbb{R}} \cap \underline{\underline{T}}^{+}$.

Tout ordre compatible définit une décomposition d'Iwasawa

$$\underline{\underline{K}} = \underline{\underline{M}} + \underline{\underline{T}}^{-} + \underline{\underline{N}} \tag{2.3.1}$$

où $\underline{\underline{N}}$ est la somme des sous-espaces de poids positifs de $\underline{\underline{T}}^{-}$ dans $\underline{\underline{K}}$. Réciproquement, toute décomposition du type (2.3.1) peut être obtenue à partir d'un ordre compatible, et deux telles décompositions sont conjuguées par le groupe de Weyl W_1 de $\underline{\underline{T}}^{-}$ dans $\underline{\underline{M}}$.

2.4 - Pour tout $\underline{\underline{K}}$-module irréductible de dimension finie V, le sous-espace $V^{\underline{\underline{N}}}$, (voir (2.3.1)), est irréductible pour l'action du centralisateur de $\underline{\underline{T}}^{-}$ dans $\underline{\underline{K}}$.

Si $V^{\underline{\underline{M}}} \neq 0$, pour tout couple $(v,w) \in V^{\underline{\underline{M}}} \times V^{\underline{\underline{N}}}$ de vecteurs non nuls, et toute forme bilinéaire $\underline{\underline{K}}$-invariante $\underline{b}$ sur V :

$$\underline{b}(v,w) \neq 0 . \tag{2.4.1}$$

En particulier

$$\dim V^{\underline{\underline{N}}} = \dim V^{\underline{\underline{M}}} = 1 . \tag{2.4.2}$$

Enfin, le poids de $\underline{\underline{T}}^{-}$ dans $V^{\underline{\underline{N}}}$ est de multiplicité un dans V.

2.5 - On suppose désormais que la paire $(\underline{\underline{G}},\underline{\underline{K}})$ n'est pas hermitienne symétrique ou, de façon équivalente, que le $\underline{\underline{K}}$-module $\underline{\underline{P}}$ est irréductible. En particulier, le centre de $\underline{\underline{K}}$ est trivial et $\underline{\underline{K}}$ est semi-simple. On suppose en outre qu'il existe un élément $H_{\circ}$ de $\underline{\underline{P}}_{\circ}$ dont le centralisateur dans $\underline{\underline{K}}_{\circ}$ est exactement $\underline{\underline{M}}_{\circ}$. En particulier, le $\underline{\underline{K}}$-module irréductible $\underline{\underline{P}}$ est $\underline{\underline{M}}$-sphérique.

2.6 - On dira qu'un poids ν de $\underline{\underline{T}}^{-}$ dans $\underline{\underline{P}}$ est extrémal, s'il existe un ordre compatible pour lequel ν est le poids dominant de $\underline{\underline{T}}^{-}$ dans $\underline{\underline{P}}$. On note $\pm\nu_1,\ldots,\pm\nu_\ell$ l'ensemble des poids extrémaux. Lorsqu'il n'y aura aucune ambiguïté sur l'ordre compatible choisi on supposera que

$$\nu_1 > \nu_2 > \ldots > \nu_\ell > 0 . \tag{2.6.1}$$

Remarquons enfin que les poids extrémaux sont W_1-conjugués.

2.7 Proposition.

Pour tout couple $1 \leq i \neq j \leq \ell$

$$[\underline{\underline{P}}^{\nu_i}, \underline{\underline{P}}^{\pm\nu_j}] = 0$$

Démonstration.

Il existe un ordre admissible pour lequel ν_i (respect. ν_j) est le plus grand poids de $\underline{\underline{T}}^-$ dans $\underline{\underline{P}}$. Par conjugaison par le groupe de Weyl de $(\underline{\underline{K}},\underline{\underline{T}}^-)$ dans $\underline{\underline{M}}$, on peut supposer $i = 1$ (ou $j = 1$!).

Si $X \in \underline{\underline{P}}^{\nu}$ $\overline{X} \in \underline{\underline{P}}^{-\nu}$; le sous-espace :

$$\tilde{\underline{\underline{P}}}' = \underline{\underline{P}}^{\nu}1 + \underline{\underline{P}}^{-\nu}1 + \underline{\underline{P}}^{\nu}j + \underline{\underline{P}}^{-\nu}j ,$$

est stable par θ et l'application $X \to \overline{X}$. il engendre une sous-algèbre $\hat{\underline{\underline{G}}}$ de $\underline{\underline{G}}$ ayant les mêmes propriétés, donc complexifiée de $\hat{\underline{\underline{G}}}_\circ = \hat{\underline{\underline{G}}} \cap \underline{\underline{G}}_\circ$ qui est θ-stable, donc réductive dans $\underline{\underline{G}}_\circ$. De plus

$$\hat{\underline{\underline{G}}}_\circ = \hat{\underline{\underline{K}}}_\circ + \hat{\underline{\underline{P}}}_\circ \quad \text{où} \quad \hat{\underline{\underline{K}}}_\circ = \hat{\underline{\underline{G}}}' \cap \underline{\underline{K}}_\circ \quad \text{et} \quad \hat{\underline{\underline{P}}}_\circ = \hat{\underline{\underline{G}}} \cap \underline{\underline{P}}_\circ$$

est une décomposition de Cartan de $\hat{\underline{\underline{G}}}_\circ$. Comme, de plus (voir (voir 2.1) :

$$[[\underline{\underline{P}}^{\nu}, \underline{\underline{P}}^{-\nu}], \underline{\underline{P}}^{\nu}] = \underline{\underline{P}}^{\nu},$$

pour tout poids de $\underline{\underline{T}}$ dans $\underline{\underline{P}}$, $\tilde{\underline{\underline{P}}}'$ est contenu dans $[\hat{\underline{\underline{G}}}, \hat{\underline{\underline{G}}}]$ et donc $\hat{\underline{\underline{G}}}' = [\hat{\underline{\underline{G}}}', \hat{\underline{\underline{G}}}']$ est semi-simple. La projection orthogonale H' de H sur $\hat{\underline{\underline{P}}}_\circ$ commute avec $\hat{\underline{\underline{M}}}_\circ = \underline{\underline{M}} \cap \hat{\underline{\underline{K}}}_\circ$; elle est non nulle car :

$$B(H', \underline{\underline{P}}^{\nu_1}) = B(H, \underline{\underline{P}}^{\nu_1}) \neq 0 .$$

Enfin, on a les relations suivantes :

$$[\underline{\underline{P}}^{\nu_i}, \underline{\underline{P}}^{\nu_i}] \subset \underline{\underline{K}}^{2\nu_i} = \{0\} , \qquad (1 \leq i \leq \ell) .$$

Supposons que $[\underline{\underline{P}}^{\nu_1}, \underline{\underline{P}}^{\nu_j}] \subset \underline{\underline{K}}^{\nu_1+\nu_j}$, (respect. $[\underline{\underline{P}}^{\nu_1}, \underline{\underline{P}}^{-\nu_j}] \subset \underline{\underline{K}}^{\nu_1-\nu_j}$), soit non nul. Cela signifie que $\alpha = \nu_1 + \nu_j$, (respect. $\beta = \nu_1 - \nu_j$) est une racine de $(\underline{\underline{K}},\underline{\underline{T}})$ et que la TDS correspondante

$$\underline{\underline{S}}_\alpha = \underline{\underline{K}}^{\alpha} + \underline{\underline{K}}^{-\alpha} + [\underline{\underline{K}}^{\alpha}, \underline{\underline{K}}^{-\alpha}] \quad (\text{respect. } \underline{\underline{S}}_\beta = \underline{\underline{K}}^{\beta} + \underline{\underline{K}}^{-\beta} + [\underline{\underline{K}}^{\beta}, \underline{\underline{K}}^{-\beta}])$$

est contenue dans $\hat{\underline{\underline{K}}}$.

$$\left.\begin{array}{l}[\underline{\underline{K}}^{\alpha}, \underline{\underline{K}}^{\beta}] \subset \underline{\underline{K}}^{2\nu_1} = 0 \\ [\underline{\underline{K}}^{\alpha}, \underline{\underline{K}}^{-\beta}] \subset \underline{\underline{K}}^{2\nu_j} = 0\end{array}\right\} \Rightarrow [\underline{\underline{S}}_{\alpha}, \underline{\underline{S}}_{\beta}] = 0$$

D'autre part :

$$[\underline{\underline{K}}^{\alpha}, \underline{\underline{P}}^{\nu_1}] \subset \underline{\underline{P}}^{2\nu_1+\nu_j} = 0 \qquad \text{car} \quad 2\nu_1 + \nu_j > \nu_1 ,$$

$$[\underline{\underline{K}}^{\alpha}, \underline{\underline{P}}^{\nu_j}] \subset \underline{\underline{P}}^{\nu_1+2\nu_j} = 0 \qquad \text{car} \quad \nu_1 + 2\nu_j > \nu_1 ,$$

$$[\underline{\underline{K}}^{\alpha}, \underline{\underline{P}}^{-\nu_1}] \subset \underline{\underline{P}}^{\nu_j} ,$$

$$[\underline{\underline{K}}^{\alpha}, \underline{\underline{P}}^{-\nu_j}] \subset \underline{\underline{P}}^{\nu_1} , \qquad \text{soit} \quad [\underline{\underline{S}}^{\alpha}, \tilde{\underline{\underline{P}}}'] \subset \tilde{\underline{\underline{P}}}'$$

De même

$$[\underline{\underline{S}}_{\beta}, \tilde{\underline{\underline{P}}}] \subset \tilde{\underline{\underline{P}}}'$$

Cela signifie que : $\mathbb{C}H_{\nu_1} + \mathbb{C}H_{\nu_j} + \underline{\underline{S}}_{\alpha} + \underline{\underline{S}}_{\beta} + \tilde{\underline{\underline{P}}}'$ est une sous-algèbre contenant $\tilde{\underline{\underline{P}}}'$ soit

$$\hat{\underline{\underline{G}}} = \underline{\underline{S}}_{\alpha} + \underline{\underline{S}}_{\beta} + \tilde{\underline{\underline{P}}}' + \mathbb{C}H_{\nu_1} + \mathbb{C}H_{\nu_j}$$

$$\hat{\underline{\underline{K}}} = \underline{\underline{S}}_{\alpha} + \underline{\underline{S}}_{\beta} + \mathbb{C}H_{\nu_1} + \mathbb{C}H_{\nu_j}$$

$$\tilde{\underline{\underline{P}}}' = \hat{\underline{\underline{P}}}$$

Si $\underline{\underline{S}}_{\alpha} \neq \{0\}$, la représentation de $\underline{\underline{S}}_{\alpha}$ dans $\hat{\underline{\underline{P}}}$ est décomposée en deux représentations de dimension deux de $\hat{\underline{\underline{P}}}$. Tout élément non nul et semi-simple de $\underline{\underline{S}}_{\alpha}$ agit donc par un opérateur bijectif de $\underline{\underline{S}}_{\alpha}$. Or, $\alpha = \nu_1 + \nu_j$ étant nul sur $\underline{\underline{T}}^{+}$, $\sigma\alpha = -\alpha$; $\underline{\underline{S}}_{\alpha}$ est stable par σ et l'application $X \to \overline{X}$. Si $X \in \underline{\underline{K}}^{\alpha}$ est non nul $X + \sigma X \in \underline{\underline{M}} \cap \underline{\underline{S}}_{\alpha}$ est non nul. Cela signifie que $\underline{\underline{M}}_{\circ} \cap S_{\alpha}$ est non nul et formé d'éléments semi-simples. Comme $[H', \underline{\underline{M}}_{\circ} \cap \underline{\underline{S}}_{\alpha}] = 0$ on aboutit à une contradiction. Soit $\underline{\underline{S}}_{\alpha} = 0$ et de même $\underline{\underline{S}}_{\beta} = 0$.

2.8 - Corollaire.

Les vecteurs $H_{\nu_j} \in [\underline{\underline{P}}^{\nu_j}, \underline{\underline{P}}^{-\nu_j}]$

tels que : $\nu_j(H_{\nu_j}) = 2$, $(1 \leqslant j \leqslant \ell)$, sont deux à deux orthogonaux et engendrent le $\mathbb{R}$-espace $\underline{\underline{T}}^-_{\mathbb{R}} = \sqrt{-1}\ \underline{\underline{T}}_\circ$

Démonstration.

D'après l'étude des représentations des TDS (voir Kostant [5])

$$\underline{\underline{S}}_j = \underline{\underline{P}}^{\nu_j} + \underline{\underline{P}}^{-\nu_j} + [\underline{\underline{P}}^{\nu_j}, \underline{\underline{P}}^{-\nu_j}]$$

on a :

$$\nu_j(H_{\nu_k}) = 0 \qquad (1 \leqslant j \neq k \leqslant \ell) .$$

et les vecteurs H_{ν_j} sont deux à deux orthogonaux.

Si $\cap \operatorname{Ker} \nu_j \cap \underline{\underline{T}}^- \neq 0$, on pourrait former une base admissible avec un vecteur $H_1 \in \cap \operatorname{Ker} \nu_j \cap \underline{\underline{T}}^-_{\mathbb{R}}$, et tout poids de $\underline{\underline{P}}$ non nul en H_1 définirait un poids $\nu > \nu_j$ $(1 \leqslant j \leqslant \ell)$. On a donc $[H_1, \underline{\underline{P}}] = 0$ soit $[H_1, \underline{\underline{G}}] = 0$. C'est exclu.

2.9 - Etant données deux formes linéaires λ et μ sur $\underline{\underline{T}}$, on écrira $\lambda \sim \mu$ pour exprimer que λ et μ ont même restriction à $\underline{\underline{T}}^-$.

2.10 - Proposition.

Pour toute racine α de $(\underline{\underline{K}}, \underline{\underline{T}})$, on est dans l'un des cas suivants :

a) Soit $\alpha \sim 0$.

b) Soit $\alpha \sim a\nu_j$, où $a = \pm\frac{1}{2}, \pm 1$.

c) Soit $\alpha \sim b\,(\nu_i \pm \nu_j)$ où $b = \pm\frac{1}{2}$.

Démonstration.

Si on n'est pas dans le cas a) , la restriction β' de β à $\underline{\underline{T}}^-$ définit une symétrie de Weyl $\underline{\underline{s}}$ de $(\underline{\underline{K}},\underline{\underline{T}}^-)$ qui ne se réduit pas à l'identité et permute les poids extrémaux de $\underline{\underline{T}}^-$. En particulier $\underline{\underline{s}}$ n'étant pas l'identité :

$$\exists\,(i,j) \in [1,\ell] \qquad \underline{\underline{s}}\nu_i = \varepsilon\nu_j \quad , \ (\varepsilon = \pm 1) \ ,$$

avec $\varepsilon = -1$ si $i = j$. Dans ce cas :

$$\varepsilon\nu_j = \underline{\underline{s}}\,\nu_i = \nu_i + c\beta' \quad , \text{ où } \ c \neq 0 \ ,$$

$$\beta \sim \beta' = -\frac{1}{c}(\nu_i - \varepsilon\nu_j) \quad ,$$

Par conjugaison des poids extrémaux on se ramène au cas $i = 1$ et $\beta > 0$.

Introduisons une base canonique $\{H,X,Y\}$ de la TDS :

$$\underline{\underline{S}} = \underline{\underline{P}}^{\nu_1} + \underline{\underline{P}}^{-\nu_1} + [\,\underline{\underline{P}}^{\nu_1},\underline{\underline{P}}^{-\nu_1}] \, .$$

Comme :

$$[\,X,\underline{\underline{K}}^{\beta}] \subset \underline{\underline{P}}^{\nu_1+\beta} = \{0\} \ , \text{ car } \ \nu_1 + \beta > \nu_1 \ ,$$

l'espace $\underline{\underline{K}}^{\beta}$ est formé de vecteurs dominants pour une représentation irréductible de dimension finie de $\underline{\underline{S}}$. Le poids dominant est β et donc

$$m = \beta(H) \in \mathbb{N} \ ,$$

<u>Premier cas</u> : $j = 1 \qquad \beta \sim a\,\nu_1$ avec $a > 0$:

$$m = \beta(H) = a\,\nu_1(H) = 2a \in \mathbb{N}^* \quad ,$$

$$0 \neq [\,Y,\underline{\underline{K}}^{\beta}] \leqslant \underline{\underline{P}}^{\beta-\nu_1} \quad \Rightarrow \ \beta - \nu_1 \leqslant \nu_1$$

$$\Rightarrow \ a \leqslant 2$$

. Si $a = 2$: $0 \neq [\,Y,\underline{\underline{K}}^{\beta}] \subset \underline{\underline{P}}^{\beta-\nu_1} = \underline{\underline{P}}^{\nu_1}$,

$0 \neq [\,X,[\,Y,\underline{\underline{K}}^{\beta}]] \subset [\,\underline{\underline{P}}^{\nu_1},\underline{\underline{P}}^{\nu_1}] = 0$. C'est exclu .

. Si $a = \frac{3}{2}$: $0 \neq (\mathrm{ad}Y)^3\,\underline{\underline{K}}^{\beta} \subset \underline{\underline{P}}^{\beta-3\nu_1} = \underline{\underline{P}}^{-\frac{3}{2}\nu_1} = \{0\}$, car $\frac{3}{2}\nu_1 > \nu_1$.

Donc $a \leqslant 1$ et $2a \in \mathbb{N}^*$

<u>Deuxième cas</u> : $j \neq 1$. $\beta' = b(\nu_1 + \varepsilon\nu_j)$, avec $\varepsilon = \pm 1$ et $b > 0$,

$$m = \beta(H) = b\,\nu_1(H) = 2b \in \mathbb{N}^* \ ;$$

. Si $m \geq 2$: $0 \neq (\mathrm{ad}\,Y)^2\ \underline{\underline{K}}^{\beta} \subset \underline{\underline{K}}^{\beta - 2\nu_1}$,

$$\beta'' = \beta - 2\nu_1 \sim (b-2)\,\nu_1 + b\varepsilon\nu_j \ ,$$

On a alors : Soit $b = 2 \Rightarrow \beta'' \sim b\varepsilon\nu_j \Rightarrow |b| \leq 1$ d'après le cas 1 .

C'est exclu.

Soit $b \neq 2$ dans ce cas, le résultat sur β appliqué à β' prouve que

$$|b-2| = |b|$$
$$b-2 = -b$$
$$b = 1$$, cas à éliminer comme précédemment.

Dans le cas $m = 2b \leq 1$ on a également le résultat cherché.

3. <u>Construction de certaines paires symétriques.</u>

3.1 - Pour ν poids de $\underline{\underline{T}}$ dans $\underline{\underline{P}}$, on désigne par H_ν l'élément de

$[\underline{\underline{P}}^{\nu}, \underline{\underline{P}}^{-\nu}] \subset \underline{\underline{T}}$ tel que :

$$\nu(H_\nu) = 2 \ ,$$

et on définit, pour $\varepsilon = \pm 1$, l'élément de $\underline{\underline{T}}^-_{\mathbb{R}}$:

$$^{\varepsilon}H = \begin{cases} H_{\nu_1} & \text{si } \ell = 1 \\ H_{\nu_1} + \dots + H_{\nu_{\ell-1}} + \varepsilon H_{\nu_\ell} & \text{si } \ell > 1 \ . \end{cases}$$

3.2 - <u>Lemme</u> :

Pour tout poids ν de $\underline{\underline{T}}$ dans $\underline{\underline{G}}$, $\nu(^{\varepsilon}H) \in \mathbb{Z}$ et ,

$$|\nu(^{\varepsilon}H)| \leq 2 \ .$$

Démonstration

On traite le cas $\ell \geq 2$: si ν est poids de $\underline{\underline{T}}$ dans $\underline{\underline{P}}$ tel que $a = |\nu(^{\varepsilon}H)| > 2$. Il existe une base $\{H_1, \dots, H_\ell\}$ de $\underline{\underline{T}}^-_{\mathbb{R}}$ telle que :

$$H_1 \in \mathbb{R}\ ^{\varepsilon}H \ , \quad \text{et} \quad \nu(H_1) = a > 2 \ .$$

Alors, pour l'ordre lexicographique correspondant :

$$\nu > \pm \nu_j \quad , \quad (1 \leqslant j \leqslant \ell) .$$

ce qui contredit le choix des ν_j .

Si ν est poids de $\underline{\underline{T}}$ dans $\underline{\underline{K}}$, on est dans l'un des cas suivants :

. Soit $\nu \sim 0$ ce qui implique $\nu({}^{\varepsilon}H) = 0$

. Soit $\nu \sim a\,\nu_j$, $(a = \pm \frac{1}{2}, \pm 1)$

Dans ce cas $|\nu({}^{\varepsilon}H)| = |a|\,\nu_j(H_{\nu_j}) = 2\,|a|$.

. Soit $\nu \sim b(\gamma_i \pm \gamma_j)$, $(b = \pm \frac{1}{2})$

Dans ce cas $\nu({}^{\varepsilon}H) = |b|\,(2 \pm 2) = \{\ 0 \text{ ou } 4\,|b|\}$

Enfin, chaque H_{ν_j} étant l'élément semi-simple d'une TDS , ses valeurs propres sont des entiers.

$$\forall_j \in [1,\ell] \qquad \nu(H_{\nu_j}) \in \mathbb{Z}$$

et donc : $\nu({}^{\varepsilon}H) \in \mathbb{Z}$.

3.3 - Définition.

On désigne par

$$
\begin{aligned}
{}^{\varepsilon}\underline{\underline{P}}^{\pm} &= \{X \in \underline{\underline{P}} \ / \ [{}^{\varepsilon}H, X] = \pm 2X\} \quad , \\
{}^{\varepsilon}\underline{\underline{\widetilde{K}}} &= \{X \in \underline{\underline{K}} \ / \ [{}^{\varepsilon}H, X] = 0\} \quad , \\
{}^{\varepsilon}\underline{\underline{P}} &= {}^{\varepsilon}\underline{\underline{P}}^{+} + {}^{\varepsilon}\underline{\underline{P}}^{-} \\
{}^{\varepsilon}\underline{\underline{\widetilde{G}}} &= {}^{\varepsilon}\underline{\underline{\widetilde{K}}} + {}^{\varepsilon}\underline{\underline{P}} \ .
\end{aligned}
$$

3.4 - Lemme.

${}^{\varepsilon}\underline{\underline{\widetilde{G}}}$ est une sous algèbre de $\underline{\underline{G}}$ stable par θ et $\eta : X \to \overline{X}$.

Elle est réductive dans $\underline{\underline{G}}$ et, si ${}^{\varepsilon}\underline{\underline{G}} = [{}^{\varepsilon}\underline{\underline{\widetilde{G}}}, {}^{\varepsilon}\underline{\underline{\widetilde{G}}}]$ désigne son algèbre dérivée :

$${}^{\varepsilon}\underline{\underline{G}}_{\circ} = {}^{\varepsilon}\underline{\underline{K}}_{\circ} + {}^{\varepsilon}\underline{\underline{P}}_{\circ} \ , \quad ({}^{\varepsilon}\underline{\underline{G}}_{\circ} = {}^{\varepsilon}\underline{\underline{G}} \cap \underline{\underline{G}}_{\circ} \ , \ {}^{\varepsilon}\underline{\underline{K}}_{\circ} = {}^{\varepsilon}\underline{\underline{G}} \cap \underline{\underline{K}}_{\circ}, {}^{\varepsilon}\underline{\underline{P}}_{\circ} = \underline{\underline{G}}_{\circ} \cap \underline{\underline{P}}_{\circ})$$

est une décomposition de Cartan de ${}^{\varepsilon}\underline{\underline{G}}_{\circ}$. De plus ${}^{\varepsilon}\underline{\underline{\widetilde{K}}}$ et ${}^{\varepsilon}\underline{\underline{K}}$ sont σ-stables.

Démonstration :

Puisque : $[{}^{\varepsilon}H, X] = aX$, $[{}^{\varepsilon}H, Y] = bY$ $\Rightarrow [{}^{\varepsilon}H, (X,Y)] = (a+b)[X,Y]$

on a trivialement :

$$0 = [{}^{\varepsilon}\underline{\underline{P}}^{+}, {}^{\varepsilon}\underline{\underline{P}}^{+}] = [{}^{\varepsilon}\underline{\underline{P}}^{-}, {}^{\varepsilon}\underline{\underline{P}}^{-}] \quad ,$$
$$[{}^{\varepsilon}\underline{\underline{P}}^{+}, {}^{\varepsilon}\underline{\underline{P}}^{-}] \subset {}^{\varepsilon}\tilde{K} \quad ;$$
$$[{}^{\varepsilon}\underline{\underline{\tilde{K}}}, {}^{\varepsilon}\underline{\underline{P}}^{\pm}] \subset {}^{\varepsilon}\underline{\underline{P}}^{\pm} \quad .$$

ce qui implique que ${}^{\varepsilon}\underline{\underline{\tilde{G}}}$ est une sous-algèbre de $\underline{\underline{G}}$.

Comme $\eta(\sqrt{-1}\,{}^{\varepsilon}H) = \theta(\sqrt{-1}\,{}^{\varepsilon}H) = \sqrt{-1}\,{}^{\varepsilon}H$, ${}^{\varepsilon}\underline{\underline{\tilde{K}}}$ et ${}^{\varepsilon}\underline{\underline{P}}$ sont stables par et θ . Il en est de même de ${}^{\varepsilon}\underline{\underline{\tilde{G}}}$ qui est la complexifiée de l'algèbre :

$${}^{\varepsilon}\underline{\underline{\tilde{G}}}_{\circ} = {}^{\varepsilon}\underline{\underline{\tilde{G}}} \cap \underline{\underline{G}}_{\circ} \quad ,$$

Cette algèbre étant θ-stable, elle est réductive dans $\underline{\underline{G}}_{\circ}$.

Enfin :

$${}^{\varepsilon}H \in {}^{\varepsilon}\underline{\underline{\tilde{K}}} \quad , \text{ et } \quad [{}^{\varepsilon}H, {}^{\varepsilon}\underline{\underline{P}}^{\pm}] = {}^{\varepsilon}\underline{\underline{P}}^{\pm} \quad ,$$

entraînent ${}^{\varepsilon}\underline{\underline{P}}^{\pm} \subset {}^{\varepsilon}\underline{\underline{G}}$. Le reste est évident.

3.5 - <u>Remarque</u>:

Pour $1 \leq j \leq \ell$, $\underline{\underline{P}}^{\pm\nu_j} \subset {}^{\varepsilon}\underline{\underline{P}}$, et donc : $H_{\nu_j} \in [\underline{\underline{P}}^{+\nu_j}, \underline{\underline{P}}^{-\nu_j}] \subset {}^{\varepsilon}\underline{\underline{K}}$.

En particulier $\underline{\underline{T}}^{-} \subset {}^{\varepsilon}\underline{\underline{K}}$.

3.6 - <u>Proposition</u>.

a) La paire $({}^{\varepsilon}\underline{\underline{G}}, {}^{\varepsilon}\underline{\underline{K}})$ est hermitienne symétrique tubulaire irréductible de rang ℓ .

b) ${}^{\varepsilon}\underline{\underline{T}} = {}^{\varepsilon}\underline{\underline{K}} \cap \underline{\underline{T}}$ est une sous-algèbre de Cartan compacte de ${}^{\varepsilon}\underline{\underline{G}}$.

c) Si $\{H_1,\dots,H_\ell\}$ et $\{H_{\ell+1},\dots,H_q\}$ sont des IR-bases de $\underline{\underline{T}}^{-}_{R}$ et $\underline{\underline{T}}^{+}_{\mathbb{R}}$ telles que $H_1 = {}^{\varepsilon}H$, $H_j = H_{\nu_{j-1}}$, $2 \leq j \leq \ell$ pour l'ordre lexicographique correspondant

$$\{\nu_1,\dots,\nu_{\ell-1}, \varepsilon\nu_\ell\} \quad , (\ell \geq 2) \quad \text{forme un système fortement}$$

orthogonal maximal de racines positives imaginaires non compactes tel que $\nu_1 > \nu_2 > \dots > \varepsilon\nu_\ell$.

Démonstration.

Pour cet ordre, le poids dominant de $\underline{\underline{P}}$ est ν_1 . Soit $X \in {}^{\varepsilon}\underline{\underline{P}}^+$ de poids ν ; comme $\underline{\underline{P}}$ est $\underline{\underline{K}}$ - irréductible :

$$\exists a \in \underline{\underline{U}}(\underline{\underline{K}}) \qquad X = a.X_{\nu_1}$$

On peut supposer que a est combinaison linéaire de produits de vecteurs dont les poids positifs ont pour somme $\nu_1 - \nu$.

Comme $\nu_1({}^{\varepsilon}H) = \nu({}^{\varepsilon}H) = 2$ par construction, tous les monômes qui figurent dans a ont pour poids 0 , donc formés de vecteurs de ${}^{\varepsilon}\underline{\underline{\tilde{K}}}$ et même de ${}^{\varepsilon}\underline{\underline{K}}$

On a donc :

$$^{\varepsilon}\underline{\underline{P}}^+ = \underline{\underline{U}}({}^{\varepsilon}\underline{\underline{K}})\, X_{\nu_1}$$

X_{ν_1} étant dominant, le ${}^{\varepsilon}\underline{\underline{K}}$-module ${}^{\varepsilon}\underline{\underline{P}}^+$ est irréductible.

La paire $({}^{\varepsilon}\underline{\underline{G}}, {}^{\varepsilon}\underline{\underline{K}})$ est donc hermitienne symétrique irréductible et ${}^{\varepsilon}\underline{\underline{T}} = {}^{\varepsilon}\underline{\underline{G}} \cap \underline{\underline{T}}$ est une sous-algèbre de Cartan compacte de ${}^{\varepsilon}\underline{\underline{G}}$.

Enfin ${}^{\varepsilon}H = H_{\nu_1} + \ldots + \varepsilon H_{\nu_\ell}$ appartenant au centre de $\underline{\underline{K}}$, la paire est de type tubulaire.

Dans une paire symétrique de type tubulaire, tout système orthogonal de racines imaginaires non compactes est fortement orthogonal. Si $\{\nu_1, \ldots, \nu_{\ell-1}, \varepsilon\nu_\ell\}$ n'était pas maximal, on pourrait le plonger dans un système orthogonal maximal et il existerait une racine imaginaire non compacte ν telle que

$$\nu({}^{\varepsilon}\underline{\underline{H}}) = 0$$

ce qui est exclu, car ν est non nulle sur le centre de ${}^{\varepsilon}\underline{\underline{K}}$.

3.7 - <u>Corollaire</u> :

a) Quels que soient les entiers distincts i et j de $[1,\ell]$ il existe un poids ν de $\underline{\underline{T}}$ dans ${}^{\pm\varepsilon}\underline{\underline{K}}$ tel que

$$\nu \sim \frac{1}{2}(\nu_i - \nu_j)$$

b) Quels que soient les entiers i et j de $[1,\ell]$ il existe un poids de $\underline{\underline{T}}$ dans ${}^{\pm\varepsilon}\underline{\underline{P}}$ tel que

$$\nu \sim \frac{1}{2}(\nu_i + \nu_j)$$

Démonstration

Les poids de $^{\varepsilon}\underline{\underline{T}} = \underline{\underline{T}} \cap {}^{\varepsilon}\underline{\underline{K}} \supseteq T^{-}$ dans $\underline{\underline{G}}$ s'obtiennent par restriction de poids de $\underline{\underline{T}}$ dans $\underline{\underline{G}}$. Or, les poids de $^{\varepsilon}\underline{\underline{T}}$ dans $^{\varepsilon}\underline{\underline{G}}$ vérifient les conditions ci-dessus (voir Harish-Chandra [3]).

3.8 - Lemme.

On conserve l'ordre lexicographique précédemment défini et on note $\Delta^{+}_{\underline{\underline{K}}}$ (respect. $\Delta^{+}_{\varepsilon\tilde{\underline{\underline{K}}}}$, respect. $\Delta^{+}_{\varepsilon_{\underline{\underline{K}}}}$) les sytèmes de racines positives de $(\underline{\underline{K}},\underline{\underline{T}})$ (respect. $(^{\varepsilon}\tilde{\underline{\underline{K}}},\underline{\underline{T}})$, respect. $(^{\varepsilon}\underline{\underline{K}}, {}^{\varepsilon}\underline{\underline{T}})$).

Alors, pour tout $\alpha \in \Delta^{+}_{\underline{\underline{K}}}$:

$$\underline{\underline{K}}^{\alpha} \not\subset {}^{\varepsilon}\underline{\underline{K}} \Rightarrow [\underline{\underline{K}}^{\alpha}, {}^{\varepsilon}\underline{\underline{P}}^{+}] = 0 .$$

En particulier, tout vecteur de poids $\Delta^{+}_{\varepsilon_{\underline{\underline{K}}}}$ dominant de $S(^{\varepsilon}\underline{\underline{P}}^{+})$ est vecteur de poids $\Delta^{+}_{\underline{\underline{K}}}$ dominant.

Démonstration.

Par hypothèse, on est dans l'un des cas suivants :

. Soit $\alpha(^{\varepsilon}H) = 0$ alors $\underline{\underline{K}}^{\alpha}$ est contenu dans un idéal (compact) de $^{\varepsilon}\underline{\underline{G}}$, et donc $[^{\varepsilon}\underline{\underline{K}}, {}^{\varepsilon}\underline{\underline{P}}^{+}] = 0$.

. Soit $\alpha(^{\varepsilon}H) > 0$ et tout vecteur de $[\underline{\underline{K}}^{\alpha}, {}^{\varepsilon}\underline{\underline{P}}^{+}]$ est de poids $2 + \alpha(^{\varepsilon}H) > 2$ pour $ad\,{}^{\varepsilon}H$, et donc $[^{\varepsilon}\underline{\underline{K}}, {}^{\varepsilon}\underline{\underline{P}}^{+}] = 0$.

Si $a \in S(^{\varepsilon}\underline{\underline{P}}^{+})$ est un vecteur de poids $\Delta^{+}_{\varepsilon_{\underline{\underline{K}}}}$ dominant ν pour $^{\varepsilon}\underline{\underline{T}}$

. a est vecteur de poids ν pour $\underline{\underline{T}}$ où $\nu = 0$ sur le supplémentaire orthogonal de $^{\varepsilon}\underline{\underline{T}}$ dans $\underline{\underline{T}}$, supplémentaire qui est contenu dans un idéal compact de $^{\varepsilon}\tilde{\underline{\underline{G}}}$.

. Pour $\alpha \in \Delta^{+}_{\underline{\underline{K}}}$:

- Soit $\alpha \in \Delta_{\varepsilon_{\underline{\underline{K}}}}$, et $\underline{\underline{K}}^{\alpha}.a = 0$ par hypothèse.

- Soit $\alpha \in \Delta^{+}_{\underline{\underline{K}}} - \Delta^{+}_{\varepsilon_{\underline{\underline{K}}}}$, et $\underline{\underline{K}}^{\alpha}.a = 0$ d'après la remarque 3.12 .

c) Si, par exemple, $X = [{}^{\varepsilon}H'', \underline{\underline{P}}^{\nu_1}] \subset \underline{\underline{K}}^{\nu_1}$ est non nul, $X + \sigma X \in \underline{\underline{M}}$ commute avec $H_\circ$ et la composante $[{}^{\varepsilon}H'', X]$ de poids ν_1 de $[H_\circ, X + \sigma X]$ est seule nulle. Cela signifie que

$$(\mathrm{ad}^{\varepsilon}H'')^2 (\underline{\underline{P}}^{\nu_1}) = 0 \quad \text{, et donc que :}$$

(3.9.13) $$[{}^{\varepsilon}H'', \underline{\underline{P}}^{\nu_1}] = 0$$

3.10 - Corollaire.

a) L'algèbre ${}^{\varepsilon}\underline{\underline{M}} = {}^{\varepsilon}\underline{\underline{K}} \cap \underline{\underline{M}}$ est le centralisateur de ${}^{\varepsilon}H'$ dans ${}^{\varepsilon}\underline{\underline{K}}$.

b) La paire $({}^{\varepsilon}\underline{\underline{K}}, {}^{\varepsilon}\underline{\underline{M}})$ définit une décomposition de Cartan de ${}^{\varepsilon}\underline{\underline{K}}$.

On note ${}^{\varepsilon}\underline{\underline{Q}} = \underline{\underline{Q}} \cap {}^{\varepsilon}\underline{\underline{K}}$.

c) ${}^{\varepsilon}H''$ centralise ${}^{\varepsilon}\underline{\underline{G}}$.

Démonstration.

a) Le Si ${}^{\varepsilon}\underline{\underline{P}}^{\perp}$ est l'orthogonal de ${}^{\varepsilon}\underline{\underline{P}}$ dans $\underline{\underline{P}}$, la décomposition

(3.10.1) $$\underline{\underline{P}} = {}^{\varepsilon}\underline{\underline{P}} + {}^{\varepsilon}\underline{\underline{P}}^{\perp}$$

est ${}^{\varepsilon}\underline{\underline{M}}$ stable, ce qui se traduit par :

(3.10.2) $$[{}^{\varepsilon}\underline{\underline{M}}, H] = 0 \Rightarrow [{}^{\varepsilon}\underline{\underline{M}}, {}^{\varepsilon}H'] = 0 \;;$$

${}^{\varepsilon}\underline{\underline{M}}$ est contenue dans le centralisateur de ${}^{\varepsilon}H'$ dans ${}^{\varepsilon}\underline{\underline{K}}$. L'inclusion inverse résultera de c).

b) C'est une conséquence immédiate de la stabilité de ${}^{\varepsilon}\underline{\underline{K}}$ par σ.

c) Le centralisateur de ${}^{\varepsilon}H''$ dans ${}^{\varepsilon}\underline{\underline{K}}$ contient :

$$\cdot \; H_{\nu_j} \in [\underline{\underline{P}}^{\nu_j}, \underline{\underline{P}}^{-\nu_j}], \quad (1 \leqslant j \leqslant \ell)$$

d'après (3.9.7.) et donc $\underline{\underline{T}}^-$;

$$\cdot \; {}^{\varepsilon}\underline{\underline{M}} \quad \text{d'après (3.10 a))}$$

Comme $\underline{\underline{T}}^-$ est un sous-espace de Cartan de ${}^{\varepsilon}\underline{\underline{Q}}$, ce centralisateur contient ${}^{\varepsilon}\underline{\underline{K}}$ tout entier. Or, dans la paire symétrique $({}^{\varepsilon}\underline{\underline{G}}, {}^{\varepsilon}\underline{\underline{K}})$, le sous-espace du centralisateur de ${}^{\varepsilon}H''$ dans ${}^{\varepsilon}\underline{\underline{P}}$,

(2.10.3) $$\Sigma_{j=1}^{\ell} (\underline{\underline{P}}^{\nu_j} + \underline{\underline{P}}^{-\nu_j})$$

contient un sous-espace de Cartan de ${}^{\varepsilon}\underline{\underline{P}}$. Autrement dit, le centralisateur de ${}^{\varepsilon}H''$ contient ${}^{\varepsilon}\underline{\underline{G}}$.

3.11 Remarque : Si $\nu \neq 0$, $H_{\nu} \in {}^{\varepsilon}\underline{\underline{P}} \cap {}^{-\varepsilon}\underline{\underline{P}}$ (voir (2.9.12)) , si $\nu = 0$ $H_{\nu} \in {}^{\varepsilon}\underline{\underline{P}}^{\perp} \cap {}^{-\varepsilon}\underline{\underline{P}}^{\perp}$; il en résulte (et on notera H', H'' ces éléments) :

(3.11.1) $$ {}^{\varepsilon}H' = {}^{-\varepsilon}H' \, , \; {}^{\varepsilon}H'' = {}^{-\varepsilon}H'' \, , \; {}^{\varepsilon}\underline{\underline{M}} = {}^{-\varepsilon}\underline{\underline{M}} \; . $$

3.12 Remarque : Toute valeur propre λ de ${}^{\varepsilon}H$ dans $S(\underline{\underline{P}})_m$ vérifie

(3.12.1) $$ |\lambda| \leqslant 2m \; , $$

l'égalité n'étant possible que si le vecteur propre correspondant appartient à

(3.12.2) $$ S({}^{\varepsilon}\underline{\underline{P}}^{+})_m \cup S({}^{\varepsilon}\underline{\underline{P}}^{-})_m \; . $$

3.13 - Lemme.

Si V est un ${}^{\varepsilon}\underline{\underline{K}}$ - type non trivial de $S({}^{\varepsilon}\underline{\underline{P}}^{\pm})_m$ on a :

(3.13.1) $$ B(V, H'^{m}) \neq 0 \; . $$

Démonstration

C'est un résultat classique sur les fonctions holomorphes sur un domaine symétrique irréductible (voir Schmidt [6]).

4. Détermination de certains $\underline{\underline{K}}$ - types de $U(\underline{\underline{G}})$.

4.1 - On fixe sur $\underline{\underline{T}}_{\mathbb{R}}$ un ordre compatible associé à la base ordonnée $\{{}^{\varepsilon}H, H_{\nu_1}, \ldots, H_{\nu_{\ell-1}}\}$ de $\underline{\underline{T}}_{\mathbb{R}}^{-}$. On note Φ_k^{+} le système de racines positives correspondant de l'ensemble Φ_k des racines de $(\underline{\underline{K}}, \underline{\underline{T}})$, (voir 2.1) . Tout K-module irréductible V est déterminé, à un isomorphisme près, par son poids Φ_k^{+} dominant λ . On notera V_{λ} une réalisation d'un tel module et v_{λ} un vecteur non nul de poids λ de V_{λ}

4.2 - Soit M le centralisateur de $H_\circ$ dans K . On va déterminer les K-modules irréductibles V_λ dans l'algèbre enveloppante $\underline{\underline{U}}(\underline{\underline{G}})$ de $\underline{\underline{G}}$ pour lesquels $V_\lambda^M \neq \{0\}$. On distinguera deux cas :

Cas I . Il existe $\nu \in \Phi_k$, $i \in [1,\ell]$ et $a \in \mathbb{R}^*$ tels que

(4.2.1) $\nu \sim a\,\nu_i$

Cas II . Les autres cas.

Remarquons que, dans le cas I on peut, en utilisant une conjugaison par le groupe de Weyl $\underline{\underline{W}}$, supposer $i \in [1,\ell]$ arbitraire (voir 1.4 et 2.7) . Dans le cas II , quels que soient i et j distincts dans $[1,\ell]$, il existe un élément ν de Φ_k tel que

(4.2.2) $\nu \sim \frac{1}{2}(\nu_i + \nu_j)$.

Car, dans le cas contraire , $\Sigma_{j=1}^{\ell} H_{\nu_j}$ est central dans $\underline{\underline{K}}$.

Enfin, dans le cas I , on supposera que le nombre $\varepsilon = \pm 1$ introduit en 3.2.1 et 4.1 vaut 1 .

4.3 - Lemme.

Soit $V_\lambda \subset \underline{\underline{U}}(\underline{\underline{G}})$ un K-module irréductible de poids Φ_k^+ dominant λ tel que

$$V_\lambda^M \neq \{0\} .$$

Alors

(4.3.1) $\lambda = n_1\nu_1 + \ldots + n_\ell \nu_\ell$

avec $(n_1,\ldots,n_\ell) \in \mathbb{Z}$ et

(4.3.2) $n_1 \geqslant n_2 \geqslant \ldots \geqslant n_{\ell-1} \geqslant |n_\ell|$

et dans le cas I

(4.3.3) $n_\ell \geqslant 0$

Démonstration.

Le poids Φ_k^+ - dominant λ de tout K-module M-sphérique V_λ est nul sur $\underline{\underline{T}}^-$ (voir § 2) . Un calcul classique (voir Kostant [5]) montre que

(4.3.4) $m_i = \exp \sqrt{-1}\,\pi \operatorname{ad} H_{\nu_i} \in M$, $(1 \leqslant i \leqslant \ell)$.

L'action de m_i sur le vecteur dominant implique (voir [5])

(4.3.5) $$1 = \exp \sqrt{-1}\,\pi\,\lambda(H_{\nu_i}) = \exp 2\pi\sqrt{-1}\, n_i \quad ,$$

c'est-à-dire

(4.3.6) $$n_i \in \mathbb{Z} \quad , \quad (1 \leqslant i \leqslant \ell) \; .$$

On peut (en raisonnant par exemple dans la paire hermitienne symétrique $({}^{\varepsilon}\underline{\underline{G}}, {}^{\varepsilon}\underline{\underline{K}})$) trouver, pour tout $i \in [1,\ell-1]$, $\nu \in \Phi_k^+$ et $a > 0$ tels que :

(4.3.7) $$\nu \sim a(\nu_i - \nu_{i+1}) \; ,$$

de telle sorte que

(4.3.8) $$\lambda(H_\nu) \in \mathbb{R}_*^+(n_i - n_{i+1}) \geqslant 0 \quad .$$

De même, dans le cas I , on peut choisir $\nu \in \Phi_k^+$ et $b > 0$ tels que (voir 4.2.1)

(4.3.9) $$\nu \sim \nu_\ell \; ,$$

et, dans le cas II , choisir $\nu \in \Phi_k^+$ et $c > 0$ tel que (voir 4.2.2) :

(4.3.10) $$\nu \sim c(\nu_{\ell-1} + \nu_\ell) \; ,$$

et la dernière assertion du Lemme résulte alors de la relation :

(4.3.11) $$\lambda(H_\nu) \geqslant 0 \; .$$

4.4 Proposition.

Soit λ une forme linéaire sur $\underline{\underline{T}}$ vérifiant les conditions de 4.3 et V un K-module irréductible de poids Φ_k^+ - dominant λ . Alors, avec les notations de 3.12, pour tout entier naturel $m \in \mathbb{N}$

a) Si V figure dans $S(\underline{\underline{P}})_m$

(4.4.1) $$n_1 + \ldots + n_{\ell-1} + |n_\ell| \geqslant m$$

b) Si

(4.4.2) $$n_1 + \ldots + n_{\ell-1} + |n_\ell| = m$$

et si on choisit $\varepsilon = \pm 1$ dans (3.1) de telle sorte que

(4.4.3) $$\varepsilon\, n_\ell \geqslant 0$$

le K-module V figure avec multiplicité un dans $S(\underline{\underline{P}})_m$, son vecteur Φ_k^+ - dominant appartient à $S({}^{\varepsilon}\underline{\underline{P}}_+)_m$ et

(4.4.4) $$V^M \neq \{0\}$$

Démonstration :

a) Si $v \in S(\underline{\underline{P}})_m$ est Φ_k^+ -dominant de poids λ

(4.4.5) $\qquad {}^{\varepsilon}H.v = 2(n_1 + \ldots + n_{\ell-1} + \varepsilon\, n_\ell)\, v$

et (4.4.1) résulte de 2.11.1.

b) De même, la relation (4.4.2) n'est possible que si $v \in S({}^{\varepsilon}\underline{\underline{P}}_+)_m$. L'existence d'un tel vecteur Φ_k^+ -dominant est, d'après le lemme 3.8 , ramenée à l'étude du cas de la paire hermitienne symétrique $({}^{\varepsilon}\underline{\underline{G}}, {}^{\varepsilon}\underline{\underline{K}})$. Dans ce cas, la réduction du ${}^{\varepsilon}\underline{\underline{K}}$-module module $S({}^{\varepsilon}\underline{\underline{P}}_{\pm})$ est bien connue (voir Schmid [6] ou Guillemonat [1]) . Le K-module V figure dans $S(\underline{\underline{P}})_m$ avec la multiplicité un et, avec les notations de 3.12.2 .

$$B(V \cap S({}^{\varepsilon}\underline{\underline{P}}_+)_m , H^m) = B(V \cap S({}^{\varepsilon}\underline{\underline{P}}_+)_m , ({}^{\varepsilon}H')^m) \neq 0$$

Si on fixe $w \in V \cap S({}^{\varepsilon}\underline{\underline{P}}_+)_m$ tel que

(4.4.6) $\qquad B(w , H^m) \neq 0$

et une mesure de Haar normalisée sur le groupe compact M , l'élément

(4.4.7) $\qquad w_o = \int_M m\, w\, dm$

appartient à V^M et vérifie

(4.4.8) $\qquad B(w_o , H^m) = B(w , H^m) \neq 0$.

4.5 Le résultat de Guillemonat est plus précis (voir [1] 3.8) . Si λ et μ sont les poids Φ_k^+ -dominants de deux K-modules irréductibles V et W de $S(\underline{\underline{P}})$ vérifiant les hypothèses de 4.4b et tels que

$$\lambda = \mu + \nu$$

avec $\nu \in \{ \nu_1, \ldots, \nu_{\ell-1}, \varepsilon\nu_\ell \}$ tout vecteur Φ_k^+ -dominant v de V vérifie

(4.5.1) $\qquad v \in (W \cap S({}^{\varepsilon}\underline{\underline{P}}_+))\, {}^{\varepsilon}\underline{\underline{P}}_+$

Comme l'algèbre ${}^{\varepsilon}\underline{\underline{P}}_+$ est abélienne, si $\hat{V}_\lambda$ (respect. $\hat{V}_\mu$) est l'image de V (respect. W) dans $\underline{\underline{U}}(\underline{\underline{G}})$ de V (respect. W) par l'application canonique

$$F : S(\underline{\underline{G}}) \to \underline{\underline{U}}(\underline{\underline{G}}) \quad , \text{ on a :}$$

(4.5.2) $\qquad \hat{V}_\lambda \subset \hat{V}_\mu\, \underline{\underline{P}}$

On notera $\hat{v}_\lambda$ un vecteur Φ_k^+ - dominant non nul fixé de $\hat{V}_\lambda$.

4.6 Soient $\underline{\underline{B}}'_\circ$ l'algèbre dérivée de la sous-algèbre parabolique maximale $\underline{\underline{B}}_\circ$ de l'algèbre semi-simple $\underline{\underline{G}}_\circ$ et $\underline{\underline{A}}_\circ$ le complément orthogonal dans $\underline{\underline{B}}_\circ \cap \underline{\underline{P}}_\circ$. L'espace $\underline{\underline{A}}_\circ$ est de dimension un et on peut choisir $\underline{\underline{H}}_\circ \in \underline{\underline{A}}_\circ$ de telle sorte que $\underline{\underline{B}}_\circ$ soit la somme des sous-espaces propres correspondant aux valeurs propres positives ou nulles de ad $\underline{\underline{H}}_\circ$. Dans le cas ou $(\underline{\underline{G}}_\circ,\underline{\underline{K}}_\circ)$ est une paire hermitienne symétrique de type tubulaire, il existe une classe de conjugaison de sous-algèbre parabolique maximale et une seule telle que la paire $(\underline{\underline{K}}_\circ,\underline{\underline{K}}_\circ \cap \underline{\underline{B}}_\circ)$ soit symétrique. C'est précisément cette situation que nous avons étudiée en détail dans nos travaux antérieurs (voir [1] et [2]).

4.7 Reprenons les notations et hypothèses du § 2 . Si $\underline{\underline{B}}_\circ$ est la somme des sous-espaces propres de $\underline{\underline{G}}_\circ$ correspondant aux valeurs propres positives ou nulles de ad $\underline{\underline{H}}_\circ$, comme :

$$\underline{\underline{M}}_\circ = \underline{\underline{B}}_\circ \cap \underline{\underline{K}}_\circ \tag{4.7.1}$$

définit une paire symétrique $(\underline{\underline{K}}_\circ,\underline{\underline{M}}_\circ)$, $\underline{\underline{B}}_\circ$ est une sous-algèbre parabolique maximale de $\underline{\underline{G}}_\circ$ telle que, avec les notations de 4.5

$$\underline{\underline{A}}_\circ = \mathbb{R}\, \underline{\underline{H}}_\circ . \tag{4.7.2}$$

Enfin

$$^{\varepsilon}\underline{\underline{B}}_\circ = {}^{\varepsilon}\underline{\underline{G}}_\circ \cap \underline{\underline{B}}_\circ . \tag{4.7.3}$$

et la sous-algèbre parabolique de $^{\varepsilon}\underline{\underline{G}}_\circ$ associée par le même procédé à l'élément $^{\varepsilon}H'$ défini en 3.9 . Dans ce cas, la paire $({}^{\varepsilon}\underline{\underline{G}}\,,{}^{\varepsilon}\underline{\underline{K}})$ étant hermitienne symétrique de type tubulaire, les résultats de [1] et [2] s'appliquent sans modification.

On note

$$^{\varepsilon}\underline{\underline{A}}_\circ = \mathbb{R}\, {}^{\varepsilon}H' \tag{4.7.4}$$

$$^{\varepsilon}\underline{\underline{B}}'_\circ = [\, {}^{\varepsilon}\underline{\underline{B}}_\circ\,,\, {}^{\varepsilon}\underline{\underline{B}}_\circ] \tag{4.7.5}$$

4.8 Définissons comme en [2] § 3 , les applications $\zeta : \underline{\underline{U}}(\underline{\underline{G}}) \to S(\underline{\underline{A}})$ et $^{\varepsilon}\zeta : \underline{\underline{U}}({}^{\varepsilon}\underline{\underline{G}}) \to S({}^{\varepsilon}\underline{\underline{A}})$ par les relations :

$$a = \zeta(a) \mod(\underline{\underline{K}}\, U(\underline{\underline{G}}) + \underline{\underline{U}}(\underline{\underline{G}})\underline{\underline{B}}') \quad , \quad (a \in \underline{\underline{U}}(\underline{\underline{G}})) , \tag{4.8.1}$$

$$a = {}^{\varepsilon}\zeta(a) \mod({}^{\varepsilon}\underline{\underline{K}}\, \underline{\underline{U}}({}^{\varepsilon}\underline{\underline{G}}) + \underline{\underline{U}}({}^{\varepsilon}\underline{\underline{G}})\, {}^{\varepsilon}\underline{\underline{B}}') , \quad (a \in \underline{\underline{U}}({}^{\varepsilon}\underline{\underline{G}})) . \tag{4.8.2}$$

Dans l'espace abélien

(4.8.3) $\mathfrak{C}_\circ = \mathbb{R}^{\varepsilon} H' + \mathbb{R}^{\varepsilon} H''$

contenant $\underline{\underline{A}}_\circ$ et $^{\varepsilon}\underline{\underline{A}}_\circ$, la projection orthogonale de $\mathfrak{C}_\circ$ sur $\underline{\underline{A}}_\circ = \mathbb{R} H_\circ$ se prolonge en un homomorphisme

(4.8.4) $\omega : S(\mathfrak{C}) \to S(\underline{\underline{A}})$

On vérifie sans difficulté que :

(4.8.5) $\zeta(a) = \omega(^{\varepsilon}\zeta(a))$, $(a \in \underline{\underline{U}}(^{\varepsilon}\underline{\underline{G}}))$.

4.9 Proposition :

a) Pour tout $a \in \underline{\underline{U}}(\underline{\underline{G}})$

(4.9.1) $\zeta(a) = \zeta(\int_M m.a\ dm)$

où m.a est le résultat de l'action adjointe de $m \in M \subset \mathrm{Int}\ \underline{\underline{G}}_\circ$ sur $a \in \underline{\underline{U}}(\underline{\underline{G}})$ et dm la mesure de Haar normalisée de M .

b) Pour tout K-module irréductible $V_\lambda \subset \underline{\underline{U}}(\underline{\underline{G}})$ admettant v_λ pour vecteur Φ_k^+ - dominant non nul

(4.9.2) $\zeta(V_\lambda) = \mathbb{C}\ \zeta(v_\lambda)$

c) Pour tout K-module irréductible $V_\lambda \subset \underline{\underline{U}}(\underline{\underline{G}})$

(4.9.3) $\zeta(V_\lambda \underline{\underline{U}}(\underline{\underline{G}})) \subset \zeta(V_\lambda)\ S(\underline{\underline{A}})$

Démonstration.

a) Voir [2] Lemme 3.5 .

b) Résulte de a) et du fait que (voir 2.4.1)

$$V_\lambda^M = \mathbb{C} \int_M m.v_\lambda\ dm$$

c) Voir [2] Lemme 7 .

4.10 Remarque :

En particulier si $V_\lambda \subset \underline{\underline{U}}(\underline{\underline{G}})$ est un K-module irréductible tel que $V_\lambda^M = \{0\}$, on a $\zeta(V_\lambda) = 0$.

4.11 Lemme :

Soient V_λ et V_μ deux k-modules irréductibles dans $\underline{\underline{U}}(\underline{\underline{G}})$, de poids Φ_k^+ - dominants λ et μ, de vecteurs Φ_k^+ - dominants v_λ et v_μ, vérifiant les conditions suivantes :

(4.11.1) $V_\lambda^M \neq \{0\}$ et $V_\mu^M \neq \{0\}$.

(4.11.2) $V_\lambda \subset V_\mu \underline{\underline{P}}$

Alors :

a) Il existe $i \in [1,\ell]$ et $\varepsilon' = 0$, -1 ou $+1$ tel que

(4.11.3) $\lambda = \mu + \varepsilon' \nu_i$

b) Si $\zeta(v_\mu)$ est divisible par $\zeta(\hat{v}_\mu)$ (voir 4.5) alors $\zeta(v_\lambda)$ est divisible par $\zeta(\hat{v}_\lambda)$.

Démonstration

D'après la Proposition 4.4, la relation 4.11 implique

(4.11.4) $\lambda = \Sigma_{j_\ell=1}^{\ell} n_j \nu_j$, $(n_j \in \mathbb{Z}, 1 \leq j \leq \ell)$

(4.11.5) $\mu = \Sigma_{j=1}^{\ell} m_j \nu_j$, $(m_j \in \mathbb{Z}, 1 \leq j \leq \ell)$

D'autre part, comme l'application $V_\mu \otimes \underline{\underline{P}} \to V_\mu \underline{\underline{P}}$ qui à $v \otimes X \in V_\mu \otimes \underline{\underline{P}}$ associe $vX \in \underline{\underline{U}}(\underline{\underline{G}})$ est surjective, le poids Φ_k^+ - dominant λ figure dans $V_\mu \otimes \underline{\underline{P}}$; il s'écrit donc sous la forme

(4.11.6) $\lambda = \mu + \nu$

où ν est un poids de $\underline{\underline{T}}$ dans $\underline{\underline{P}}$. Cela signifie que

(4.11.7) $\nu = \varepsilon' \nu_j$ où $j \in [1,\ell]$

C'est l'assertion a). Pour l'assertion b), remarquons tout d'abord que $\zeta(V_\mu)$ divise $\zeta(V_\lambda)$ (voir 4.9.3).

. Si $\varepsilon' = 0$, $\mu = \lambda$ et $\hat{v}_\mu = \hat{v}_\lambda$ et l'assertion b) est triviale.

. Si $\varepsilon' = -1$ $\varepsilon' \nu_i$ est négative, la relation

$$\mu = \lambda - \varepsilon' \nu_i$$

implique (voir 4.5) que $\hat{V}_\mu \subset \hat{V}_\lambda \underline{\underline{P}}$.

On en déduit (voir (4.9.3) que $\zeta(\hat{v}_\lambda)$ divise $\zeta(\hat{v}_\mu)$, donc $\zeta(v_\mu)$ et $\zeta(v_\lambda)$. Il reste à démontrer l'assertion pour $\varepsilon'\nu_j$ positive; dans ce cas (voir (4.5)) :

$$(4.11.8) \qquad \hat{V}_\lambda \subset \hat{V}_\mu \underline{\underline{P}}$$

Soient V une réalisation du K-module V_μ , v_1 un vecteur Φ_k^+ - dominant non nul de V , φ la forme linéaire sur V telle que, pour $v \in V$

$$(4.11.9) \qquad \int_M m.v_1 \, dm = \varphi(v)\, v'_1 \ , \quad 0 \neq v'_1 \in V^M \ ,$$

soit la projection M-équivariante de v sur V^M . On fixe deux K-isomorphismes $T_1 : V \to \hat{V}_\mu$ et $T_2 : V \to V_\mu$, et on note $S_i : V \otimes \underline{\underline{P}} \to \underline{\underline{U}}(\underline{\underline{G}})$ le K-morphisme défini par :

$$(4.11.10) \qquad S_i(v \otimes X) = T_i(v)\, X \quad , \quad (v \in V, X \in \underline{\underline{P}}\ , i = 1,2) \ .$$

La multiplicité du K-module V_λ dans $V \otimes \underline{\underline{P}}$ est au plus celle du poids $\varepsilon'\nu_j$ dans $\underline{\underline{P}}$, soit un . Soit

$$(4.11.11) \qquad \omega = \Sigma_{j=1}^{n} v_j \otimes X_j \quad , \quad (v_j \in V, X_j \in \underline{\underline{P}}\ , 1 \leq j \leq n) \ ,$$

un vecteur non nul de poids Φ_k^+ - dominant λ dans $V \otimes \underline{\underline{P}}$. On écrit, pour $1 \leq j \leq n$

$$(4.11.12) \qquad X_j = X'_j + X''_j + X'''_j \quad , \quad (X'_j \in \underline{\underline{K}}, X''_j \in \underline{\underline{A}}, X'''_j \in \underline{\underline{B}}')$$

de telle sorte que pour $i = 1,2$:

$$S_i(\omega) = \Sigma_{j=1}^{n} (-\operatorname{ad} X'_j\, T_i(v_j) + T_i(v_j) X''_j) \bmod (\underline{\underline{K}}\,\underline{\underline{U}}(\underline{\underline{G}}) + \underline{\underline{U}}(\underline{\underline{G}})\underline{\underline{B}}')$$

$$\zeta(S_i(\omega)) = \Sigma_{j=1}^{n} (-\zeta(T_i(X'_j . v_j)) + \zeta(T_i(v_j) X''_j) \ ,$$

$$= \xi(T_i(v'_1))\, (\Sigma_{j=1}^{n} - \varphi(X'_j . v_j) + \varphi(v_j) X''_j) \ .$$

Finalement (voir 4.9.a)) , on a, pour $i = 1,2$:

$$(4.11.13) \qquad \zeta(S_i(\omega)) = \zeta(T_i(v'_1))\, (\Sigma_{j=1}^{n} - \varphi(X'_j . v_j) + \varphi(v_j) X''_j) \ .$$

Il suffit alors de remarquer que le second facteur ne dépend que de V (et non des T_i) ; comme on peut supposer que :

(4.11.14) $\hat{v}_\mu = T_1(v_1)$ et $v_\mu = T_2(v_1)$,

(4.11.15) $\hat{v}_\lambda = S_1(\omega)$ et $v_\lambda = S_2(\omega)$,

on obtient immédiatement le résultat.

4.12 Théorème.

Soit V_λ un K-module irréductible dans $\underline{\underline{U}}(\underline{\underline{G}})$, de poids Φ_k^+ - dominant λ . On suppose que
que

(4.12.1) $V_\lambda^M \neq \{0\}$

Alors, avec les notations de 3.5 et 3.2 , $\zeta(\hat{v}_\lambda)$ divise $\zeta(v_\lambda)$.

Démonstration

On note

(4.12.2) $S(\underline{\underline{P}})_{(m)} = \sum_{j=0}^{m} S(\underline{\underline{P}})_j$.

On sait que l'application canonique (voir 3.5) de $F(S(\underline{\underline{P}}))\otimes \underline{\underline{U}}(\underline{\underline{K}})$ sur $\underline{\underline{U}}(\underline{\underline{G}})$ est un K-isomorphisme. On raisonne par récurrence sur l'entier

(4.12.3) $n(V_\lambda) = \mathrm{Inf}\ \{ m \in \mathbb{N}\ /\ V_\lambda \subset F(S(\underline{\underline{P}})_{(m)})\otimes \underline{\underline{U}}(\underline{\underline{K}})\}$

On peut alors supposer

(4.12.4) $V_\lambda \subset V_\mu \underline{\underline{P}}$

où V_μ est un K-module irréductible tel que

$$n(V_\mu) < n(V_\lambda)$$

Si $V_\mu^M = \{0\}$ on a successivement (voir 4.9.3 et 4.10).
$\zeta(v_\mu) = 0$ et $\zeta(v_\lambda) = 0$; l'assertion est évidente. Si $V_\mu^M \neq \{0\}$, l'hypothèse de récurrence permet d'appliquer le lemme 4.11 .

4.13 Appelons ε_P la multiplicité commune (d'après l'étude des paires hermitiennes symétriques) des poids de $\varepsilon_{\underline{\underline{G}}}$, soit compacts et équivalents à $\pm\frac{1}{2}(\nu_i-\nu_j)$, $1 \leq i \leq j \leq \ell$ et $\pm\frac{1}{2}(\nu_i+\varepsilon\nu_\ell)$, soit non compacts et équivalents à $\pm\frac{1}{2}(\nu_i+\nu_j)$ $1 \leq i \neq j \leq \ell-1$ et $\pm\frac{1}{2}(\nu_i-\varepsilon\nu_\ell)$, $1 \leq i < \ell-1$.

Rappelons que $[{}^{\pm\varepsilon}\underline{\underline{G}}, H''] = 0$.

Lemme .

(i) La dimension de l'espace des vecteurs de poids $\frac{1}{2}(v_i \pm v_j)$, $E^{\pm}_{i,j}$, $1 \leqslant i \neq j \leqslant \ell$ fixés est égale à $({}^{\varepsilon}p + {}^{-\varepsilon}p)$. De plus $[E_{i,j}, H''] = 0$.

(ii) Si $\ell \geqslant 3$ $\quad {}^{\varepsilon}p = {}^{-\varepsilon}p$.

Preuve.

L'espace des vecteurs de $\underline{\underline{K}}$ (resp $\underline{\underline{P}}$) de poids $v_1 + \varepsilon v_\ell$ est inclus par définition dans ${}^{-\varepsilon}\underline{\underline{K}}$ (resp ${}^{+\varepsilon}\underline{\underline{P}}$) et donc commute avec H'' . Ad $\underline{\underline{G}} - {}^{1}\underline{\underline{K}}$ contient des éléments qui stabilisent $\underline{\underline{T}}'^{-}$, permuttent les H_{ν_i} et laissent fixe H'' . (i) est donc prouvé .

(ii) L'existence pour $\ell > 1$ de racines compactes $\pm\frac{1}{2}(\nu_i \pm \nu_j)$, $1 \leqslant i \neq j \leqslant \ell$, suffit à assurer pour $\ell \geqslant 3$ la conjugaison de $\frac{1}{2}(\nu_1 - \nu_2)$ par exemple avec $\frac{1}{2}(\nu_1 + \nu_2)$ dans le dual de $\underline{\underline{T}}^{-}$.

Q.E.D.

4.14 On considére le sous-espace E engendré par les vecteurs de poids $\pm\nu_i$ de $\underline{\underline{P}}$. Il existe un sous-espace réel abélien maximal $\underline{\underline{a}}_\circ$, de dimension ℓ, tel que $[\underline{\underline{a}}_\circ, \underline{\underline{H}}_\circ] = 0$, de l'espace $E \wedge \underline{\underline{P}}_\circ$. D'après la définition de H' , $H' \in \underline{\underline{a}}_\circ$. La seule hypothèse faite précédemment sur ${}^{\circ}\underline{\underline{A}}_\circ$ était la condition $H \in {}^{\circ}\underline{\underline{A}}_\circ$; on supposera désormais $H'' \in {}^{\circ}\underline{\underline{A}}_\circ$, $\underline{\underline{a}}_\circ \subset {}^{\circ}\underline{\underline{A}}_\circ$ (donc $H \in {}^{\circ}\underline{\underline{A}}_\circ$) . Il existe un élément de Ad $\underline{\underline{g}}\ {}^{\varepsilon}\underline{\underline{G}}$, τ , d'après l'étude des paires symétriques tel que

(4.14.1) $\quad \tau\,\underline{\underline{a}} = \underline{\underline{T}}^{-}$.

Posons :

(4.14.2) $\quad E' = \mathbb{R}\,H'' + \underline{\underline{a}}_\circ$.

On déduit de (4.14.1) l'existence de vecteurs de poids sur E' : φ_i , $\frac{1}{2}(\varphi_i \pm \varphi_j)$ et l'existence d'une base de E' B': $\{H'', H_1, \ldots H_\ell\}$ tels que $1 \leqslant i \leqslant \ell$:

(4.14.3) $\quad \varphi_i(H_j) = 2\delta_{ij}$.

(4.14.4) $\varphi_i : (H") = 0$.

De tels vecteurs de poids sur E' se prolongent en une combinaison linéaire de vecteur de poids sur $°\underline{\underline{A}}_\circ$ que l'on peut supposer positifs si $i = 1$ d'après (4.14.4).

4.15 Lemme :

$$2\rho(H_{\varphi_1}) > ({}^{\varepsilon}p + {}^{-\varepsilon}p)(\ell-1)$$

Preuve : Si φ est une racine restreinte positive non nulle sur $\underline{\underline{A}}_\circ$, d'après (4.14.1) et (1.10) :

(4.15.1) $[X_{\varphi}, X_1] = 0$, X_1 étant un vecteur de poids φ_1 .

donc $\varphi(H_1) \geq 0$. Il suffit alors de considérer les racines restreintes à E' : $\frac{1}{2}(\varphi_1 \pm \varphi_i)$, $2 \leq i \leq \ell$ de multiplicité ${}^{\varepsilon}p + {}^{-\varepsilon}p$, d'après 4.13 , et φ_1 .

Q.E.D.

4.16 Soit $\overline{H}$ la projection de $\frac{H_{\varphi_1}}{2}$ sur $\mathbb{R}H_\circ$

Lemme $°\rho(\overline{H}) > \frac{{}^{\varepsilon}p + {}^{-\varepsilon}p}{4}(\ell-1)$

Preuve En fait $°\rho(\overline{H}) = \rho(\overline{H}) = \rho\left(\frac{H_{\varphi_1}}{2}\right)$

puisque $\rho(A) = 0$ si $A \in A_\circ^\perp$.

Q.E.D.

4.17 Théorème Soit $\hat{v}_\lambda \in U({}^{\varepsilon}\underline{\underline{G}})$:

(4.17.1) $$\xi(\hat{v}_\lambda) = d_\lambda \prod_{k=0}^{\ell-1} \prod_{i=0}^{n_{k-1}-1} \frac{\overline{H}}{2\ell} + k\frac{{}^{\varepsilon}p}{2} - i = p_\lambda(\overline{H})$$

$0 \neq d_\lambda \in \mathbb{C}$.

Preuve Comme $\hat{v}_\lambda \in U({}^{\varepsilon}\underline{\underline{G}})$. D'après Guillemonat [2]]

$${}^{\varepsilon}\zeta(\hat{v}_\lambda) = p_\lambda(E) ,$$

où E est un certain élément non nul de $°\underline{\underline{A}}_\circ \cap {}^{\varepsilon}\underline{\underline{G}}$. Il s'ensuit d'après

4.8 que $\zeta(\hat{v}_\lambda)$ a la forme indiquée. La normalisation de $\overline{H}$ s'obtient en remarquant que $\hat{v}_{2\nu_1} = X^2_{\nu_1} \in {}^{\varepsilon}\underline{\underline{G}} \cap {}^{-\varepsilon}\underline{\underline{G}}$. De plus, par un calcul direct dans la T.D.S $(X_{\nu_1}, H_{\nu_1}, X_{-\nu_1})$:

$$\xi(\hat{v}_{2\nu_1}) = \frac{H_{\rho_1}}{2}\left(\frac{H_{\rho_1}}{2} - 1\right) \quad \text{et donc :}$$

$$\zeta(\hat{v}_{2\nu_1}) = \frac{\overline{H}}{2}\left(\frac{\overline{H}}{2} - 1\right)$$

Q.E.D.

4.18 Théorème H_μ est irréductible si et seulement si pour tout $\underline{\underline{K}}$-module V_λ vérifiant $V_\lambda^M \neq 0$ on a :

$$(4.18.1) \qquad (\nu, P_\lambda(+\overline{H} - {}^\circ\rho(\overline{H}))P_\lambda(-\overline{H} - {}^\circ\rho(\overline{H})) \neq 0$$

Preuve C'est une conséquence directe des résultats précédents.

Q.E.D.

4.19 Il reste à déterminer pour quelles valeurs de μ, H_μ est infinitésimalement unitaire. D'après les résultats précédents :

Théorème Si H_μ , $\mu = \nu+\rho'$, on a une structure hermitienne, H_μ sera infinitésimalement unitaire si et seulement si pour tout $\underline{\underline{K}}$-module V_λ vérifiant $V_\lambda^M \neq 0$ on a

$$(4.19.1) \qquad P_\lambda(+\overline{H} - {}^\circ\rho(\overline{H}))P_\lambda(-\overline{H} - {}^\circ\rho(\overline{H})) \geqslant 0 \ .$$

Il reste à expliciter ces conditions, ce qui nous ferons pour $\ell \geqslant 3$ laissant au lecteur la discussion pour $\ell = 1$ ou $\ell = 2$.

Théorème Si $\ell \geqslant 3$ (et donc ${}^{\varepsilon}p = {}^{-\varepsilon}p = p$) et ν réel vérifie la condition (1.5.9) , H_μ est infinitésimalement unitaire si et seulement si :

$$|\nu(\overline{H})| \leqslant {}^\circ\rho(\overline{H}) - (\ell-1)\,p/2$$

ou bien

$$(4.19.2) \qquad |\nu(\overline{H})| = {}^\circ\rho(\overline{H}) - k\,p/2 \quad , \quad 0 \leqslant k \leqslant \ell-1$$

4.20 Remarque

D'après 4.16 $°\rho(\overline{H}) - (\ell-1)\, p/2 > 0$.

BIBLIOGRAPHIE.

[1] GUILLEMONAT A. - On some semi-spherical representations of an Hermitian symmetric pair of the tubular type, I, manuscripta mathematica 31, 331-361 (1980).

[2] GUILLEMONAT A. - On some semi-spherical representations of an Hermitian symmetric pair of the tubular type, II, Mathematische Annalen, 246, 93-116 (1980).

[3] HARISH-CHANDRA - Representations of semi-simple Lie groups VI, Amer. J. Math. 78, 564-628 (1956).

[4] HELGASON S. - Differential Geometry and Symmetric Spaces, Academic Press Inc. New-York, London, Smith and Eilenberg, 1962.

[5] KOSTANT B. - On the existence and irreducibility of certain series of representations, Summer School on Lie Groups and their representations. Budapest 1971, London, Adam Hilger 1975.

[6] SCHMIDT,W. - Die randwerke holomorpher Functionnen auf hermitisch symmetrischen Raümen. Inventiones Math., 9, 61-80 (1969).

*
* *

THE PLANCHEREL THEOREM FOR SEMISIMPLE LIE GROUPS WITHOUT COMPACT CARTAN SUBGROUPS

Rebecca Herb

Let G be a semisimple real Lie group with Lie algebra $\underline{g}$. If $G_{\mathbb{C}}$ is the simply connected complex Lie group with Lie algebra $\underline{g}_{\mathbb{C}}$, the complexification of $\underline{g}$, we assume that G is the connected subgroup of $G_{\mathbb{C}}$ corresponding to $\underline{g}$. Fix a maximal compact subgroup K of G with Cartan involution θ.

In [2b] the Plancherel formula for G was proved using the method of Fourier inversion of orbital integrals for the case that rank G = rank K and the root system of $\underline{g}$ is of classical type. In [2c] this proof was extended to include the root systems of exceptional type, still under the assumption that rank G = rank K. In this paper the methods used in the equal rank case, in particular the notion of a two-structure for a root system, will be extended to the case that rank $G >$ rank K.

The method of computing the Plancherel measure for G via Fourier inversion of orbital integrals was first used by Sally and Warner [4] in the case when G is of real rank one. The idea is as follows. Let H be a fundamental Cartan subgroup of G. If rank G = rank K, H is compact, while in general H is as compact as possible and is unique up to conjugacy. For $f \in C_c^\infty(G)$ and $h \in H'$, the set of regular elements of H, define

$$F_f(h) = \Delta(h) \int_{G/H} f(xhx^{-1})\, d\dot{x} \tag{1}$$

where Δ is the standard Weyl denominator factor defined in [1a] and $d\dot{x}$ is a suitably normalized G-invariant measure on the quotient. Except for the Δ factor, $F_f(h)$ is just the integral of f over the G-orbit of h.

Let D denote the differential operator on H given by $D = \prod_{\alpha \in \Phi^+} H_\alpha$. Here $\Phi^+ = \Phi^+(\underline{g}_{\mathbb{C}}, \underline{h}_{\mathbb{C}})$ is the system of positive roots for $\underline{g}_{\mathbb{C}}$ with respect to $\underline{h}_{\mathbb{C}}$, the complexified Lie algebra of H, used to define Δ and H_α is the root vector in $\underline{h}_{\mathbb{C}}$ dual to α via the Killing form. Then a theorem of Harish-Chandra [1b] states that $F_f(h;D)$ extends to a continuous function on all of H and that for all $f \in C_c^\infty(G)$

(2) $$F_f(1;D) = M_G f(1)$$

where M_G is a constant depending only on G. The idea of Sally and Warner was to obtain a Fourier inversion formula for the invariant distribution $f \mapsto F_f(h)$, $h \in H'$, and then differentiate and evaluate at $h = 1$ to obtain a Fourier inversion formula for the delta distribution $f \mapsto f(1)$. This last formula is the Plancherel Theorem. This approach avoids much of the difficult analysis used by Harish-Chandra in his proof of the Plancherel Theorem for reductive groups [1c].

For groups of real rank greater than one Fourier inversion formulas for certain averaged orbital integrals were computed before those for individual orbital integrals. In [2b,c] the Plancherel formula was obtained by differentiating these formulas. Now however, a Fourier inversion formula for F_f itself is available [2d], and it will be used directly.

The idea that made it possible to find an explicit Fourier inversion formula for groups of general rank is that of a two-structure. This tool allowed the formula to be computed for groups of any rank once formulas were known for the groups $SL(2,\mathbb{R})$ and $Sp(2,\mathbb{R})$. Two-structures are defined as follows.

Let Φ be any root system, W its Weyl group. A root system $\phi \subseteq \Phi$ is a two-structure for Φ if:

(i) all simple factors of ϕ are of type A_1 or $B_2 \cong C_2$;

(ii) if ϕ^+ is any set of positive roots for ϕ,

then $\{w \in W : w\phi^+ = \phi^+\}$ contains no elements of determinant -1. Let $T(\Phi)$ denote the set of all two-structures for Φ. Then all elements of $T(\Phi)$ are conjugate by W, and if Φ is simple, $T(\Phi)$ consists of all root systems $\phi \subseteq \Phi$ of the following types.

Φ:	A_{2n}	A_{2n+1}	B_{2n}	B_{2n+1}	C_{2n}	C_{2n+1}	D_{2n}	D_{2n+1}
ϕ:	A_1^n	A_1^{n+1}	B_2^n	$B_2^n \times B_1$	C_2^n	$C_2^n \times C_1$	A_1^{2n}	A_1^{2n}

Φ:	E_6	E_7	E_8	F_4	G_2
ϕ:	A_1^4	A_1^7	A_1^8	B_2^2	A_1^2.

Note that for $\phi \in T(\Phi)$, rank ϕ = rank Φ if and only if Φ is spanned by strongly orthogonal roots.

Let Φ^+ be a choice of positive roots for Φ. For each $\phi \in T(\Phi)$, let $\phi^+ = \phi \cap \Phi^+$. Define $W(\phi:\Phi^+) = \{\sigma \in W : \sigma\phi^+ \subseteq \Phi^+\}$ and $W_1(\phi:\Phi^+) =$

$\{\sigma \in W : \sigma\phi^+ = \phi^+\}$. Let $\varepsilon(\phi : \Phi^+) = [W_1(\phi:\Phi^+)]^{-1} \sum_{\sigma \in W(\phi:\Phi^+)} \det \sigma$. Clearly for any $\phi \in T(\Phi)$, $\sigma \in W(\phi:\Phi^+)$

$$(*) \qquad \varepsilon(\sigma\phi:\Phi^+) = \det \sigma\varepsilon(\phi:\Phi^+).$$

As a consequence, if ϕ_0 is any element of $T(\Phi)$,

$$\sum_{\phi \in T(\Phi)} \varepsilon(\phi:\Phi^+) = \sum_{\sigma \in W(\phi_0:\Phi^+)/W_1(\phi_0:\Phi^+)} \det \sigma\varepsilon(\phi_0:\Phi^+) = \varepsilon(\phi_0:\Phi^+)^2.$$

It can be proved by induction on the rank of Φ that in fact $\varepsilon(\phi:\Phi^+) = \pm 1$ for every $\phi \in T(\Phi)$. The above gives an intrinsic definition of the signs $\varepsilon(\phi:\Phi^+)$ which were defined in [2c] by picking a basepoint $\phi_0 \in T(\Phi)$ with $\varepsilon(\phi_0:\Phi^+) = 1$ and using (*) to define $\varepsilon(\phi:\Phi^+)$ for general $\phi \in T(\Phi)$.

The characters of G needed for the Fourier inversion of F_f are the tempered unitary spectrum parameterized by the dual groups of Cartan subgroups. For any θ-stable Cartan subgroup H of G, write $H_K = H \cap K$. Let $\underline{h}_p = \underline{h} \cap \underline{p}$ where $\underline{h}$ is the Lie algebra of $\underline{h}$ and $\underline{p}$ is the (-1)-eigenspace for θ in $\underline{g}$. Then $\hat{H}$, the dual of H, is parameterized by pairs (b^*,μ) where $b^* \in \hat{H}_K$ and $\mu \in \underline{h}^*_p$. the real dual of $\underline{h}_p$. To each such pair there is a corresponding tempered invariant eigendistribution $\theta(H,b^*,\mu)$ defined as in [2a]. If $b^* \in \hat{H}_K$ is regular, then $\theta(H,b^*,\mu)$ is, up to sign, the character of a unitary representation induced from a parabolic subgroup with split part $H_p = \exp(\underline{h}_p)$. If b^* is not regular, $\theta(H,b^*,\mu)$ is induced from limits of discrete series [3]. There are equivalences among the $\theta(H,b^*,\mu)$ given by the action of $W(G,H) = N_G(H)/H$ where $N_G(H)$ is the normalizer of H in G.

Now assume that H is fundamental and let $L = MH_p$ be the centralizer in G of H_p. M is a reductive group with compact Cartan subgroup H_K. Let $\mathrm{Car}(M)$ denote a full set of θ-stable representatives for M-conjugacy classes of Cartan subgroups of M. For each $J \in \mathrm{Car}(M)$, $\tilde{J} = JH_p$ is a Cartan subgroup of G. As above, write $J_K = J \cap K$ and $\underline{j}_p = \underline{j} \cap \underline{p}$ where $\underline{j}$ is the Lie algebra of J. Write $W(M,J) = N_M(J)/J$. Then, if Haar measures are normalized as in [2a], the Fourier inversion formula for F_f can be written as follows [2d].

<u>Theorem</u>. Let $f \in C_c^\infty(G)$ and let $h_k h_p \in H'$, $h_k \in H_K$, $h_p \in H_p$. Then

$$F_f(h_k h_p) = (2\pi)^{-\dim \underline{h}_p} \sum_{J \in \mathrm{Car}(M)} d_J \times$$

$$\sum_{b^*\in\hat{J}_k} \int_{\underline{h}_p} h_p^{-i\mu'} \int_{\underline{j}_p^*} \theta(\tilde{J},b^*,\mu\otimes\mu')(f) \sum_{w\in W(M,H_K)} \det wK(M,J,b^*,\mu,wh_k)\,d\mu\,d\mu'.$$

The constant d_J is given by

$$d_J = \left(\frac{i}{2}\right)^{\dim \underline{j}_p} \frac{(-1)^{r_I(J)}}{[W(M,J)][Z(\underline{h}_p)\cap Z(\underline{j}_p)]}$$

where $r_I(J)$ is half the number of pure imaginary roots of $\underline{j}$ and for any abelian subalgebra $\underline{a}$ of $\underline{p}$, $Z(\underline{a}) = \exp(i\underline{a}) \cap K$.

The terms K will be defined using two-structures. Fix $J \in \mathrm{Car}(M)$ and let $R = R_J$ be the set of strongly orthogonal singular imaginary roots of $\underline{h}_k$ corresponding to $\underline{j}$, ν the Cayley transform with respect to R satisfying $\nu(\underline{h}_k) = \underline{j}_k + i\underline{j}_p$. Let $\Phi_R = \Phi_R(\underline{m},\underline{j})$ be the set of real roots of $\underline{j}$ in $\underline{m}$, the Lie algebra of M. Pick a two-structure ϕ_R for Φ_R so that $\nu R \subseteq \phi_R$. Note that since M has a compact Cartan subgroup, Φ_R is spanned by strongly orthogonal roots and rank ϕ_R = rank $\Phi_R = \dim \underline{j}_p$. Let $\Phi_R^+ = \Phi_R \cap \nu\Phi^+(\underline{g}_{\mathbb{C}},\underline{h}_{\mathbb{C}})$.

Decompose $\phi_R = \phi_1 \cup \ldots \cup \phi_k$ where the ϕ_j, $1 \le j \le k$, are simple. Write $\underline{j}_p = \underline{a}_1 \oplus \ldots \oplus \underline{a}_k$ where $\underline{a}_j$ is spanned by the root vectors H_α, $\alpha \in \phi_j$. Then for any $h_k \in H_K$, νh_k can be decomposed as $\nu h_k = j_0 a_1 \ldots a_k$ where $j_0 \in J_K$ and $a_j \in \exp(i\underline{a}_j)$, $1 \le j \le k$. This decomposition is not unique. Let $E_J = \{(a_1,\ldots,a_k) : a_j \in \exp(i\underline{a}_j)$ and $\prod_{j=1}^{k} a_j = 1.\}$ For $\mu \in \underline{j}_p^*$ and $b^* \in \hat{J}_K$, let $\mu_j \in \underline{a}_j^*$ be the restriction of μ to $\underline{a}_j$ and b_j^* the restriction of b^* to $Z(\underline{a}_j)$. Then

$$K(M,J,b^*,\mu,h_K) = \varepsilon(\phi_R:\Phi_R^+)[T(\Phi_R)][E_J]^{-1}\,\overline{b^*(j_0)} \prod_{j=1}^{k} K(\phi_j,b_j^*,\mu_j,a_j).$$

The K factors corresponding to simple root systems of type A_1 and $B_2 \cong C_2$ are given explicitly in [2d] and these somewhat complicated formulas will not be reproduced here. Note that if J is the split Cartan subgroup of $SL(2,\mathbb{R})$ or $Sp(2,\mathbb{R})$, then $\phi_R = \Phi_R$ is of type A_1 or C_2 respectively and $K(M,J,b^*,\mu,h_K) = K(\phi_R,b_1^*,\mu_1,a_1)$. Thus the factors appearing for arbitrary groups are the products of factors occuring for split Cartan subgroups of $SL(2,\mathbb{R})$ and $Sp(2,\mathbb{R})$.

The formula for $F_f(h)$ in the theorem can be differentiated just as in [2b,c]. There are extra terms in each $K(\phi_j)$ factor for which ϕ_j is of type B_2 which do not occur in [2b,c] where averaged orbital integrals were used. But these terms cancel out when differentiated and so can be ignored. Thus

$$(3)\quad \lim_{h_k h_p \to 1} D[h_p^{-i\mu'} \sum_{w\in W(M,H_K)} \det w\, K(M,J,b^*,\mu,wh_k)] =$$

$$[W(M,H_K)][T(\Phi_R)]\varepsilon(\phi_R:\Phi^+)[E_J]^{-1}(-1)^{r_I(J)} \prod_{\alpha\in R_J} 2/\|\alpha\| \quad p(b^*,\mu,\mu')$$

where

$$p(b^*,\mu,\mu') = \prod_{\alpha\in\Phi^+(\underline{g}_{\mathbb{C}},\tilde{\underline{j}}_{\mathbb{C}})} (\alpha,\log b^*+i\mu+i\mu') \prod_{\alpha\in\phi_R^+} \frac{\cosh \pi\mu_\alpha - \xi_M \otimes b^*(\gamma_\alpha)}{\sinh \pi\mu_\alpha}.$$

Here $\mu_\alpha = \mu(H_\alpha)$, $\gamma_\alpha = \exp(\pi i H_\alpha)$, and $\xi_M(\gamma_\alpha) = \exp(\pi i \rho_M(H_\alpha))$ where $\rho_M = \frac{1}{2}\sum_{\alpha\in\Phi_R^+} \alpha$.

Up to this point, the fact that rank G > rank K makes little difference as we have been working primarily inside the equal rank group M. In order to obtain the Plancherel measure in the form given by Harish-Chandra in [1c] it is necessary to switch our attention to the group G itself.

As before let $\tilde{J} = JH_p$, $\tilde{\underline{j}}$ its Lie algebra. When H is not compact, $\tilde{\Phi}_R = \Phi_R(\underline{g},\tilde{\underline{j}})$, the set of real roots of $\tilde{\underline{j}}$ in $\underline{g}$, is usually of higher rank than $\Phi_R = \Phi_R(\underline{m},\underline{j})$ and is not spanned by strongly orthogonal roots. In fact, two-structures for Φ_R will also be two-structures for $\tilde{\Phi}_R$, and if $\tilde{W}_R$ denotes the Weyl group for $\tilde{\Phi}_R$, $T(\tilde{\Phi}_R) = \{\sigma\phi_R : \sigma\in\tilde{W}_R\}$ where $\phi_R \in T(\Phi_R)$. Since $\tilde{W}_R$ can be regarded as a subgroup of $W(G,\tilde{J})$, it is possible by changing variables under the integral to obtain

$$(4)\quad \varepsilon(\phi_R:\Phi_R^+)\sum_{b^*\in\hat{J}_K} \int_{\underline{h}_p^*}\int_{\underline{j}_p^*} \theta(\tilde{J},b^*,\mu\otimes\mu')(f)\,p(b^*,\mu,\mu')\,d\mu\, d\mu' =$$

$$[T(\tilde{\Phi}_R)]^{-1} \sum_{b^*\in\hat{J}_K} \int_{\tilde{\underline{j}}_p^*} \theta(\tilde{J},b^*,\mu)(f) \sum_{\phi\in T(\tilde{\Phi}_R)} \varepsilon(\phi:\tilde{\Phi}_R^+)\,\tilde{p}(b^*,\mu,\phi^+)\,d\mu \quad \text{where}$$

$$\tilde{p}(b^*,\mu,\phi^+) = \prod_{\alpha\in\Phi^+(\underline{g}_{\mathbb{C}},\tilde{\underline{j}}_{\mathbb{C}})} (\alpha,\log b^*+i\mu) \prod_{\alpha\in\phi^+} \frac{\cosh \pi\mu_\alpha - \xi_G\otimes b^*(\gamma_\alpha)}{\sinh \pi\mu_\alpha}.$$

Here $\xi_G(\gamma_\alpha) = \exp(\pi i \rho_G(\gamma_\alpha))$ where $\rho_G = \frac{1}{2}\sum_{\alpha\in\tilde{\Phi}_R^+}\alpha$.

An elementary proof of the following lemma was given by M. Vergne in [5].

<u>Lemma</u>. $$\sum_{\phi\in T(\tilde{\Phi}_R)} \varepsilon(\phi:\tilde{\Phi}_R^+) \prod_{\alpha\in\phi^+} \frac{\cosh \pi\mu_\alpha - \xi_G\otimes b^*(\gamma_\alpha)}{\sinh \pi\mu_\alpha} = \prod_{\alpha\in\tilde{\Phi}_R^+} \frac{\cosh \pi\mu_\alpha - \xi_G\otimes b^*(\gamma_\alpha)}{\sinh \pi\mu_\alpha}.$$

Combining (2), (3), and (4) with the theorem and lemma, we can write

$$(5)\quad f(1) = M_G^{-1}(2\pi)^{-\dim \underline{h}_p}[W(M,H_K)] \sum_{J\in Car(M)} d_{\tilde{J}}$$

$$\times \sum_{b^*\in \hat{J}_K} \int_{\tilde{\underline{j}}_p^*} \theta(\tilde{J},b^*,\mu)(f)p(\tilde{J},b^*,\mu)d\mu \qquad \text{where}$$

$$p(\tilde{J},b^*,\mu) = \prod_{\alpha\in\Phi^+(\underline{g}_{\mathbb{C}},\tilde{\underline{j}}_{\mathbb{C}})} (\alpha, \log b^*+i\mu) \prod_{\alpha\in\tilde{\Phi}_R^+} \frac{\cosh \pi\mu_\alpha - \xi_G \otimes b^*(\gamma_\alpha)}{\sinh \pi\mu_\alpha} .$$

The constant M_G which comes from (2) is given by $M_G = (-1)^q(2\pi)^r$ where $r = [\Phi^+(\underline{g}_{\mathbb{C}},\underline{h}_{\mathbb{C}})]$ and $q = \frac{1}{2}(\dim G/K - \text{rank } G + \text{rank } K)$. Finally
$$d_{\tilde{J}} = (\tfrac{i}{2})^{\dim \underline{j}_p}[W(M,J)]^{-1}[Z(\underline{h}_p)\cap Z(\underline{j}_p)]^{-1}[E_J]^{-1}[T(\Phi_R)][T(\tilde{\Phi}_R)]^{-1}\prod_{\alpha\in R_J} 2/\|\alpha\|.$$

It is possible to simplify significantly the constants occuring in (5). Let Car(G) denote a complete set of representatives for G-conjugacy classes of Cartan subgroups of G. Every element of Car(G) can be represented as $\tilde{J}$ for some $J \in Car(M)$. However there may be distinct elements $J_1, J_2 \in Car(M)$ with $\tilde{J}_1$ conjugate to $\tilde{J}_2$ in G. Let $Car(M,J) = \{J_1 \in Car(M) : \tilde{J}_1$ is conjugate to $\tilde{J}$ in $G\}$. Then

$$\frac{[Car(M,J)][W(M,H_K)][T(\Phi_R)]}{[W(M,J)][T(\tilde{\Phi}_R)]} = \frac{[W(G,H)]}{[W(G,\tilde{J})]} .$$

If $b^* \in \hat{J}_K$ is singular, $\prod_{\alpha\in\Phi^+(\underline{g}_{\mathbb{C}},\tilde{\underline{j}}_{\mathbb{C}})} (\alpha, \log b^* + i\mu) = 0$ for every $\mu \in \tilde{\underline{j}}_p^*$. Let $\hat{J}_K'$ denote the set of regular elements in $\hat{J}_K$. For $b^* \in \hat{J}_K'$, let $T(\tilde{J},b^*,\mu) = \pm\theta(\tilde{J},b^*,\mu)$ be the actual induced character with correct sign. Then $(-1)^q(i)^{\dim \underline{j}_p}\theta(\tilde{J},b^*,\mu)(f)p(\tilde{J},b^*,\mu) = T(\tilde{J},b^*,\mu)(f)|p(\tilde{J},b^*,\mu)|$.

For any $\phi \in T(\tilde{\Phi}_R)$, let $\phi = \phi_1 \cup \ldots \cup \phi_K$ be the decomposition of ϕ into simple factors. As before let $\underline{a}_j$ be the subspace of $\tilde{\underline{j}}_p$ spanned by root vectors from ϕ_j. Let $\underline{a}_0$ be the orthogonal complement in $\tilde{\underline{j}}_p$ of $\underline{a}_1 \oplus \ldots \oplus \underline{a}_k$ with respect to the Killing form. Let $E_{\tilde{J}} = \{(a_0, a_1, \ldots, a_K) : a_j \in \exp(i\underline{a}_j)$ and $\prod_{j=0}^{k} a_j = 1\}$. Then $[E_J][Z(\underline{h}_p) \cap Z(\underline{j}_p)] = [E_{\tilde{J}}]$.

<u>Theorem</u>. Let $f \in C_c^\infty(G)$. Then $f(1) =$

$$\frac{[W(G,H)]}{(2\pi)^{r+p}} \sum_{J\in Car(G)} c_J^{-1} \sum_{b^*\in\hat{J}_K'} \int_{\underline{j}_p^*} T(J,b^*,\mu)(f)|p(J,b^*,\mu)|d\mu$$

where $p(J,b^*,\mu)$ is defined as in (5), $r = [\Phi^+(\underline{g}_{\mathbb{C}},\underline{h}_{\mathbb{C}})]$, $p = \text{rank } G -$

rank K, and $c_J = [W(G,J)][E_J] \prod_{\alpha \in R_J} \|\alpha\|$.

References

1. Harish-Chandra, a) A formula for semisimple Lie groups, Amer. J. Math., 79 (1957), 733-760. b) Some results on an invariant integral on a semisimple Lie algebra, Ann. of Math., 80 (1964), 551-593. c) Harmonic analysis on real reductive groups III, Ann. of Math., 104 (1976), 117-201.

2. R. Herb, a) Fourier inversion of invariant integrals on semisimple real Lie groups, Trans. Amer. Math. Soc., 249 (1979), 281-302. b) Fourier inversion and the Plancherel theorem for semisimple real Lie groups, Amer. J. Math., 104 (1982), 9-58. c) Fourier inversion and the Plancherel theorem, Non-commutative Harmonic Analysis and Lie Groups, Proceedings, Marseille-Luminy 1980, Lecture Notes in Math. 880, Springer-Verlag, 1981, 197-210. d) Discrete series characters and Fourier inversion on semisimple real Lie groups, to appear Trans. Amer. Math. Soc.

3. R. Herb and P. Sally, Singular invariant eigendistributions as characters in the Fourier transform of invariant distributions, J. Funct. Analysis, 33 (1979), 195-210.

4. P. Sally and G. Warner, The Fourier transform on semisimple Lie groups of real rank one, Acta Math., 131 (1973), 1-26.

5. M. Vergne, A Poisson-Plancherel formula for semisimple Lie groups, Ann. of Math., 115 (1982), 639-666.

Department of Mathematics
University of Maryland
College Park, Maryland
20742

COMPLETION FUNCTORS IN THE $\underline{O}$ CATEGORY

A. JOSEPH

1. Introduction

1.1 This work is a natural continuation of [13] whose notation based on [5] we adopt. We extend (2.9) the validity of the braid relations ([13], 3.11 (ii)) to the whole $\underline{O}$ category, generalize (3.2) the Duflo-Zelobenko 4 step exact sequence ([13], 4.7) and obtain (5.2) a significant refinement of ([12], 4.13). For this last result we view the Enright functor as a member of a family of "completion functors" on the $\underline{O}$ category which are shown to form a semi-group (4.5).

1.2 Let $\underline{g}$ be a complex semi-simple Lie algebra, $\underline{h}$ a Cartan subalgebra for $\underline{g}$, R the set of non-zero roots, B a choice of simple roots and P(R) the lattice of weights. Throughout we fix $\lambda \in \underline{h}^*$ dominant and regular, $\Lambda = \lambda + P(R)$ and as in ([13], 2.1) restrict to the full subcategory $\underline{O}_\Lambda$ of modules in $\underline{O}$ (the Bernstein-Gelfand-Gelfand category) whose $\underline{h}$ weights lie in Λ. We set $R_\lambda = \{\alpha \in R \mid (\alpha^\vee, \lambda) \in \mathbb{Z}\}$ which is a root system with Weyl group W_λ (notation [5], 1.2).

1.3 A basic aim is to determine the composition factors of the primitive quotient $U(\underline{g})/J(\mu)$: $J(\mu) = \text{Ann}\, L(\mu)$, $\mu \in \Lambda$ (notation [5], 1.2), viewed as a $U(\underline{g}) \otimes U(\underline{g})$ module. For this it is enough by translation principles to take μ regular and even of the form $\mu = w\lambda$ with $w \in W_\lambda$. It is shown how in principle the Enright functors can be used to solve this problem, though at present we have only rederived results of Duflo ([4], Prop. 6), Vogan ([14], sect. 3) and reinterpreted ([12], 4.8).

1.4 Given $M,N \in \mathrm{Ob}\ \underline{O}_\Lambda$, let $L(M,N)$ denote the "largest Harish-Chandra submodule" of $\mathrm{Hom}_{\mathbb{C}}(M,N)$ (notation, [5], 1.5). A more natural problem than 1.3 is to determine the composition factors of the $L(L(w\lambda),L(y\lambda))$: $w,y \in W_\lambda$. Our main result (5.2) interrelates these questions and in particular extends the set of $w \in W_\lambda$ for which is known that the obvious map $U(g) \to L(L(w\lambda),L(w\lambda))$ is surjective ([12], ($\underline{C}_8$) and [13], 5.6). A further related question is to determine which primitive ideals of $U(\underline{g})/\mathrm{Ann}\ M(\lambda)$ (notation [5], 1.2) are idempotent. These is good evidence that this is always true.

2. The braid relations

2.1 Let B_λ denote the set of simple roots for R_λ defined by the positive system $R_\lambda^+ := R_\lambda \cap \mathbb{N}B$. For each $\alpha \in B_\lambda$ we have defined ([13], 2.1) a functor on $\underline{O}_\Lambda$ through $C_\alpha M = L(M(s_\alpha\lambda), M) \otimes_{U(\underline{g})} M(\lambda)$. It is covariant and left exact ([13], 2.2). Moreover from the embedding $M(s_\alpha\lambda) \hookrightarrow M(\lambda)$ we have a map $M \to C_\alpha M$. We set $D_\alpha^- M = \mathrm{Im}(M \to C_\alpha M)$ and call M α-free if the map $M \to D_\alpha^- M$ is injective. For $\alpha \in B$ this is equivalent to saying that $X_{-\alpha}$ (notation [5], 1.1) acts freely on M and we remark that on the category of α-free modules C_α is just Enright's completion functor ([13]. 2.12). Given $M \in \mathrm{Ob}\ \underline{O}_\Lambda$ let $\delta(M) \in \mathrm{Ob}\ \underline{O}_\Lambda$ denotes its "$\underline{O}$ dual" (notation [5], 1.8). The functor $M \to \delta M$ is exact, contravariant and involutory. We set $D_\alpha^+ M = \delta D_\alpha^- \delta M$, and call M α-**cofree** if the map $M \to D_\alpha^+ M$ is surjective, that is if $\delta(M)$ is α-free.

2.2 Given $M \in \mathrm{Ob}\ \underline{O}_\Lambda$ we let [M] denote its image in the Grothendieck group (which is a free $\mathbb{Z}$ module with basis $[M(\mu)] : \mu \in \Lambda$). Given $\mu \in \Lambda$ dominant, let $\underline{O}_{\hat{\mu}}$ denote the full subcategory having simple factors amongst the $L(w\mu) : w \in W_\lambda$. For μ regular these subcategories are all equivalent and have Grothendieck group $\mathbb{Z}[M(w\mu)] : w \in W_\lambda$, which can be conveniently identified with $\mathbb{Z} W_\lambda$ by identifying $[M(w\mu)]$ with w. This identification gives the Grothendieck group of $\underline{O}_{\hat{\Lambda}}$ a W_λ-W_λ bimodule structure. It is straightforward to show that C_α leaves $\underline{O}_{\hat{\Lambda}}$ stable and furthermore from ([13], 3.2) and say ([6], 3.7) it is immediate that

LEMMA. - *For all* $M \in \underline{O}_{\hat{\Lambda}}$ *one has*
$[C_\alpha M] + [D_\alpha^+ M] = (1+s_\alpha)[M]$. *In particular*
$[C_\alpha M] = s_\alpha[M]$, *iff* M *is* α-*cofree.*

2.3 Let $M, N \in \text{Ob}\ \underline{O}_\Lambda$; $\alpha \in B_\lambda$. Then $(a,b) \mapsto ab$ of $L(M,N) \times L(M(s_\alpha\lambda),M) \to L(M(s_\alpha\lambda),N)$ defines a map of $U(\underline{g}) \otimes U(\underline{g})$ modules, and so a map $L(M,N) \times C_\alpha M \to C_\alpha N$ of $U(\underline{g})$ modules. Let $T_\alpha^{M,N} : L(M,N) \to L(C_\alpha M, C_\alpha N)$ be the map of $U(\underline{g}) \otimes U(\underline{g})$ modules which results.

LEMMA. - *If* M *is* α-*cofree then*
$T_\alpha^{M,N}$ *is injective*.

Take $a \in L(M,N)$. If $a \in \text{Ker}\ T_\alpha^{M,N}$, then $a(C_\alpha M) = 0$ and so $a(L(M(s_\alpha\lambda),M)) = 0$. Yet $L(M(s_\alpha\lambda),M)M(s_\alpha\lambda) = D_\alpha^+ M$ ([13], 3.7) and so $a\,(D_\alpha^+ M) = 0$. Thus if M is α-cofree $a\,M=0$, which gives $a=0$.

2.4 Identify $U(\underline{g}) \otimes U(\underline{g})$ canonically with $U(\underline{g}\times\underline{g})$. Then the map $\eta : (x,y) \to (y,x)$ of $\underline{g}\times\underline{g}$ to itself extends to an isomorphism of $U(\underline{g}) \otimes U(\underline{g})$. If V is a $U(\underline{g}) \otimes U(\underline{g})$ module we let V^η denote the module which is V as a vector space and with the action $(a,v) \to \eta(a)v$, $\forall a \in U(\underline{g}) \otimes U(\underline{g}), \forall v \in V$. Clearly $(L(N \otimes M)^*)^\eta \cong L(M \otimes N)^*$, $\forall\ M,N \in \text{Ob}\ \underline{O}_\Lambda$ and so by ([5], 1.9) we obtain

LEMMA. - $L(M,\delta(N))^\eta \cong L(N,\delta(M))$, $\forall\ M,N \in \text{Ob}\ \underline{O}_\Lambda$.

2.5 LEMMA. - For each $\alpha \in B_\lambda$ one has

$C_\alpha \delta M(\lambda) \cong \delta M(s_\alpha \lambda)$.

Indeed $L(M(\lambda), C_\alpha \delta M(\lambda)) \cong L(M(s_\alpha \lambda), \delta M(\lambda))$

$\cong L(M(\lambda) \otimes M(s_\alpha \lambda))^* = L(-\lambda, -s_\alpha \lambda)$ (notation [5], 1.5). Now for any $w \in W_\lambda$ one has ([7], 5.1(ii)) that $L(-\lambda, -w\lambda) \cong L(-w^{-1}\lambda, -\lambda)$ and so we obtain that $L(M(\lambda), C_\alpha \delta M(\lambda)) \cong L(M(\lambda), \delta M(s_\alpha \lambda))$ from which the required assertion follows.

2.6 When $N = \delta M(\lambda)$ we simply write T_α^M for $T_\alpha^{M,N}$. Now T_α^M is a map of Harish-Chandra modules and we compute the corresponding map which is also denoted by T_α^M in the equivalent category $\underline{O}_\Lambda$. For this observe that

$L(C_\alpha M, C_\alpha \delta M(\lambda)) \cong L(C_\alpha M, \delta M(s_\alpha \lambda)) \cong L(M(s_\alpha \lambda), \delta C_\alpha M)^\eta \cong L(M(\lambda), C_\alpha \delta C_\alpha M)^\eta$ whereas $L(M, \delta M(\lambda)) \cong L(M(\lambda), \delta M)^\eta$. Thus we obtain a map $T_\alpha^M : \delta M \to C_\alpha \delta C_\alpha M$ and consequently a map $\delta T_\alpha^M : \delta C_\alpha \delta C_\alpha M \to M$.

LEMMA. - The map δT_α^M is injective with image $D_\alpha^+ M$. In particular δT_α^M (and hence T_α^M) is bijective if M is α-cofree.

For any $N \in \mathrm{Ob}\, \underline{O}_\Lambda$ it follows from 3.6(i) that $C_\alpha N$ is α-free and so $\delta C_\alpha N$ is α-cofree. Applied to $N = \delta C_\alpha M$ it follows that $\mathrm{Im}\, \delta T_\alpha^M$ is a submodule of $D_\alpha^+ M$. Applied to $N=M$ it follows from 2.2 that $[\delta C_\alpha \delta C_\alpha M] = [C_\alpha \delta C_\alpha M] = s_\alpha [\delta C_\alpha M] = s_\alpha [C_\alpha M]$.

Now $(1+s_\alpha)[N] = 0 \Leftrightarrow N$ is α-finite ([13], 2.4 and for example see proof of [13], 3.2). So by ([13], 2.4) we obtain that $(1+s_\alpha)[M/D_\alpha^+ M] = 0$. Combined with 2.2 this gives that $s_\alpha [C_\alpha M] = [D_\alpha^+ M]$ and hence the lemma.

2.7 Since λ is dominant, it easily follows that $M(\lambda)$ is projective in $\underline{O}_\Lambda$ and $\delta M(\lambda)$ is injective (as is well-known). The following result is also fairly well-known.

LEMMA. - <u>Take</u> $N \in \mathrm{Ob}\, \underline{O}_\Lambda$. <u>Then for suitable</u> $E, F \in \mathcal{E}$ <u>(notation</u> [13], 2.3) <u>there exists an exact sequence</u> $0 \to N \to E \otimes \delta M(\lambda) \to F \otimes \delta M(\lambda)$.

It is enough to prove the dual statement. The latter follows from the fact that $\underline{O}_\Lambda$ has enough projectives and every projective module in $\underline{O}_\Lambda$ is a direct summand of $E \otimes M(\lambda)$ for a suitable $E \in \mathcal{E}$. However the most natural proof comes from the corresponding assertion in the Harish-Chandra category (see for example the assertion just above the proof of ([13], 1.16 (iii))) and then for example E can be taken to be a finite dimensional $\underline{k}$ stable (notation [13], 1.3) generating subspace of $L(M(\lambda), \delta N)$.

2.8 We now refine 2.3.

PROPOSITION. - <u>If</u> M <u>is</u> α<u>-cofree</u>, then $T_\alpha^{M,N}$ <u>is bijective</u>.

Assume M is α-cofree. By 2.6, $T_\alpha^{M,N}$ is bijective if $N = \delta M(\lambda)$. Tensoring with $E \in \mathcal{E}$ (which is an exact functor commuting with C_α ([13], 2.3)) it follows that $T_\alpha^{M,N}$ is bijective if $N = \delta M(\lambda) \otimes E$. For arbitrary $N \in \mathrm{Ob}\, \underline{O}_\Lambda$, take the exact sequence in the conclusion of 2.7. Since the functors $L(M,-)$ and $L(C_\alpha M,-)$ are covariant and left exact we obtain a diagram of maps with exact rows

$$\begin{array}{ccccccc}
0 \to & L(M,N) & \to & L(M,E \otimes \delta M(\lambda)) & \to & L(M,F \otimes \delta M(\lambda)) & \\
& \downarrow T_\alpha^{M,N} & & T_\alpha^{M,E\otimes\delta M(\lambda)} \downarrow & & T_\alpha^{M,E\otimes\delta M(\lambda)} \searrow & \\
0 \to & L(C_\alpha M, C_\alpha N) & \to & L(C_\alpha M, C_\alpha(E\otimes\delta M(\lambda))) & \to & & L(C_\alpha M, C_\alpha(F\otimes\delta M(\lambda)))
\end{array}$$

which in virtue of the definition of the T_α maps is easily seen to be commutative. Then diagram chasing proves the assertion.

<u>Remark</u> : One may also prove that $T_{\alpha}^{M,N}$ is surjective if N is an injective module.

2.9 Any Verma module is α-free. Yet by ([13], 2.11) $M(w\lambda)$ is α-cofree if and only if $s_{\alpha}w \leqslant w$ (Bruhat order). So if $s_{\alpha} w \leqslant w$, then by ([13], 2.5) $C_{\alpha}M(w\lambda) \cong M(s_{\alpha}w\lambda)$ and we have an isomorphism $L(M(w\lambda),M) \xrightarrow{\sim} L(M(s_{\alpha}w\lambda),C_{\alpha}M)$ for all $M \in \mathrm{Ob}\ \underline{O}_{\Lambda}$. Thus if we define for each $w \in W_{\lambda}$ the functor C_w on $\mathrm{Ob}\ \underline{O}_{\Lambda}$ through $C_wM = L(M(w^{-1}\lambda),M) \otimes_{U(\underline{g})} M(\lambda)$ we obtain the

COROLLARY. - <u>For any reduced decomposition</u>

$w = s_{\alpha_1}s_{\alpha_2} \dots s_{\alpha_k}$ <u>one has</u> $C_w = C_{\alpha_1}C_{\alpha_2} \dots C_{\alpha_k}$.

2.10 Take $\alpha,\beta \in B_{\lambda}$ distinct. It is clear that the above result implies that the pair $\{C_{\alpha},C_{\beta}\}$ satisfy the braid relation which when say $s_{\alpha}s_{\beta}s_{\alpha} = s_{\beta}s_{\alpha}s_{\beta}$, takes the form $C_{\alpha}C_{\beta}C_{\alpha} = C_{\beta}C_{\alpha}C_{\beta}$. In the case of the Enright functors defined on $U(\underline{n}^-)$ free modules, these relations were first established independently by Bouaziz ([1], Thm.1) and Deodhar ([2],). The present situation is more delicate since one has $C_{\alpha} \neq C_{\alpha}^2 = C_{\alpha}^3$ whilst the $C_{\alpha}^2 : \alpha \in B_{\lambda}$ do not satisfy the braid relations ([13], 3.15).

The fact that the $\{C_{\alpha} : \alpha \in B_{\lambda}\}$ fail to generate a "singular Hecke algebra" (c.f. [5], 3.3) can be best understood through the following result, which determines modules in $\underline{O}_{\Lambda}$ equivalent to principal series modules.

LEMMA. - <u>For all</u> $w, y \in W_{\lambda}$ <u>one has</u>

(i) $C_w\, \delta M(\lambda) \cong \delta M(w\lambda)$.

(ii) $L(-w\lambda,-y\lambda) \otimes_{U(\underline{g})} M(\lambda) \cong C_{y^{-1}}\, C_w\, \delta M(\lambda)$.

One has $L(M(\lambda),C_w\delta M(\lambda)) \cong L(M(w^{-1}\lambda),\delta M(\lambda)) \cong L(-\lambda,-w^{-1}\lambda) \cong L(-w\lambda,-\lambda) \cong L(M(\lambda),\delta M(w\lambda))$, hence (i). For (ii) recall that $L(-w\lambda,-y\lambda) = L(M(y\lambda),\delta M(w\lambda)) = L(M(\lambda), C_{y^{-1}}\, \delta M(w\lambda))$.

2.11 From 2.9 and 2.10 we see that the principal series modules can be generated from $\delta M(\lambda)$ by the action of the $\{C_\alpha : \alpha \in B_\lambda\}$. Had C_α satisfied the additional relation $C_\alpha^2 = C_\alpha$, then we would have obtained only card W_λ isomorphism classes of principal series modules, whereas the actual number is more nearly (card $W_\lambda)^2$. We call a generalized principal series module a module of the form $C_{\alpha_1} C_{\alpha_2} \dots C_{\alpha_k} \delta M(\lambda) : \alpha_1, \alpha_2, \dots, \alpha_k \in B_\lambda$. It would be interesting to study these. For example, is the number of equivalence classes of indecomposable direct summands of generalized principle series modules finite ? We leave the following results as an exercise.

LEMMA. - _Let_ M _be a generalized principle series module. Then_

(i) $[M : M(w_\lambda \lambda)] = 1.$

(ii) _Let_ N _be the smallest submodule of_ M _with_ $[N : M(w_\lambda \lambda)] = 1$. _There exists_ $w \in W_\lambda$ _such that_ $N \cong \delta M(w\lambda)$.

(iii) _If_ $C_\alpha M \cong M$, $\forall\, \alpha \in B_\lambda$, _then_ $M \cong M(\lambda)$.

(iv) _There exist_ $\alpha_1, \alpha_2, \dots, \alpha_k \in B_\lambda$ _(not necessarily distinct) such that_ M _is a submodule of_ $C_{\alpha_1}^2 C_{\alpha_2}^2 \dots C_{\alpha_k}^2 \delta M(\lambda)$.

3. _The Duflo-Zelobenko 4 step exact sequence._

3.1 Using Frobenius reciprocity it is easy to show that $[C_{y^{-1}} C_w \delta M(\lambda)] = y^{-1} w$ (in the conventions of 2.2). Unfortunately it appears that no such simple formula extends to an arbitrary generalized principle series module. To understand this better it is appropriate to generalize ([13], 4.6). For this recall ([13], 4.4) that for each $M \in \mathrm{Ob}\, \underline{O}_\lambda$ and each $\alpha \in B_\lambda$ we constructed a map $\kappa_\alpha^M : C_\alpha \delta M \to \delta(C_\alpha M)$.

LEMMA. - _There is an exact sequence_

$$0 \to D_\alpha^-(\delta M) \to C_\alpha \delta M \xrightarrow{\kappa_\alpha^M} \delta(C_\alpha M) \to D_\alpha^+ \delta M \to 0.$$

One has $\operatorname{Ker} \kappa_\alpha^M = D_\alpha^-(\delta M)$ by ([13], 4.4). On the other hand through the definition of κ_α^M ([13], 4.4) it easily follows that the maps κ_α^M, $\delta\kappa_\alpha^{\delta M}$ of $C_\alpha \delta M \to \delta C_\alpha M$ coincide. Together with our first observation, this gives the required observation.

3.2 Set $D_\alpha = D_\alpha^+ D_\alpha^- = D_\alpha^- D_\alpha^+$.

COROLLARY. - <u>There is an exact sequence</u>

$$0 \to C_\alpha M \to C_\alpha^2 M \to D_\alpha^+ M \to D_\alpha M \to 0$$

Apply 3.1 with M replaced by $\delta C_\alpha M$. This gives $0 \to D_\alpha^- C_\alpha M \to C_\alpha^2 M \to \delta C_\alpha \delta C_\alpha M \to D_\alpha^+ C_\alpha M \to 0$. Finally apply ([13], 3.6 (i), (iv)) and 2.6.

3.3 The above result refines ([13], 3.13 (ii)) and allows one to compare $C_\alpha M$ with $C_\alpha^2 M$. If M is α-cofree and the map $D_\alpha^- M \to C_\alpha M$ is surjective, then we see that $\left[C_\alpha^2 M\right] = \left[M\right] = s_\alpha\left[C_\alpha M\right]$. This can happen if M is a principle series module and was the situation describe by the Duflo-Zelobenko sequence.

4. Completion functors

4.1 The results of section 2 lead naturally to the question of whether a product of the $C_\alpha : \alpha \in B_\lambda$ is determined by its action on $\delta M(\lambda)$. For this we define a <u>completion functor</u> C on $\underline{O}_\lambda$ to be a functor on $\underline{O}_\lambda$ satisfying

(1) C <u>is covariant and left exact</u>.

(2) <u>For each</u> $E \in \mathcal{E}$, $M \in \operatorname{Ob} \underline{O}_\lambda$ <u>there exists an isomorphism</u> $\varphi_{M,E}$ <u>such that the following diagrams commute</u>

a)

$$\begin{array}{ccc} C(E\otimes M) & \xrightarrow[\sim]{\varphi_{M,E}} & E\otimes CM \\ \big\downarrow{\scriptstyle M\to N} & & \big\downarrow{\scriptstyle M\to N} \\ C(E\otimes N) & \xrightarrow[\sim]{\varphi_{N,E}} & E\otimes CN \end{array}$$

b)

$$\begin{array}{ccc} C(E\otimes F\otimes M) & \xrightarrow{\varphi_{F\otimes M,E}} & E\otimes C(F\otimes M) \\ & \searrow{\scriptstyle \varphi_{M,E\otimes F}} & \big\downarrow{\scriptstyle \varphi_{M,F}} \\ & & E\otimes F\otimes CM \end{array}$$

c)

$$\begin{array}{ccc} C(E\otimes E^{*}\otimes M) & \xrightarrow{\varphi_{M,E\otimes E^{*}}} & E\otimes E^{*}\otimes CM \\ & \searrow{\scriptstyle e\otimes e^{*}\mapsto\langle e,e^{*}\rangle} & \big\downarrow{\scriptstyle e\otimes e^{*}\mapsto\langle e,e^{*}\rangle} \\ & & CM \end{array}$$

(3) For any finite dimensional ad $\underline{g}$ submodule E of $U(\underline{g})$ the diagram

$$\begin{array}{ccc} C(E\otimes M) & \xrightarrow{e\otimes m\mapsto em\ :\ e\in E,\ m\in M} & \\ \big\downarrow{\scriptstyle \wr\ \varphi_{M,E}} & & CM \\ E\otimes CM & \xrightarrow{e\otimes m'\mapsto em'\ :\ e\in E,\ m'\in CM} & \end{array}$$

commutes.

It is immediate that the composition of any two completion functors is again a completion functor. Note that (3) implies that $\mathrm{Ann}_{Z(\underline{g})}M \subset \mathrm{Ann}_{Z(\underline{g})}CM$.

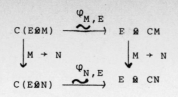

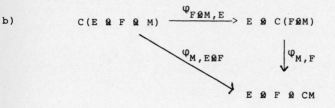

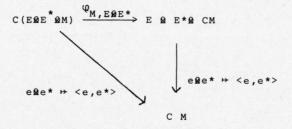

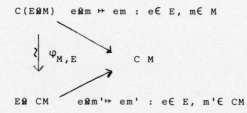

4.2 Take $N \in \mathrm{Ob}\, \underline{O}_\Lambda$ and define the covariant functor C_N on $\mathrm{Ob}\, \underline{O}_\Lambda$ through $C_N M = L(N,M) \otimes_{U(\underline{g})} M(\lambda)$. Since λ is dominant and regular, it follows from ([5], 1.16) that C_N is left exact. As in ([13], 2.3) it follows that C_N commutes with the functor $E \otimes -$. Finally using say ([13], 2.6) we find that C_N also satisfies (3). That is we have the

LEMMA. - <u>For any</u> $N \in \mathrm{Ob}\, \underline{O}_\Lambda$, C_N <u>is a completion functor</u>.

4.3 Let ε denote the covariant functor defined through $\varepsilon(N) = L(M(\lambda),N)^\eta \otimes_{U(\underline{g})} M(\lambda)$. Through ([5], 1.16) it follows that ε is exact. Obviously ε^2 is the identify functor.

LEMMA.-

(i) $\varepsilon\delta = \delta\varepsilon$.

(ii) $\varepsilon(\delta M(w\lambda)) = \delta M(w^{-1}\lambda)$.

(iii) $\varepsilon(M(w\lambda)) = M(w^{-1}\lambda)$.

(iv) $\varepsilon(L(w\lambda)) = L(w^{-1}\lambda)$.

(i) We define δ on the category of Harish-Chandra modules $\underline{H}_{\hat{\lambda}}$ (notation [5], 1.16) as in ([13], 2.8). In ([13],2.8) we defined for each $N \in \mathrm{Ob}\, \underline{O}_\Lambda$ a map $L(M(\lambda),\delta N) \to \delta(L(M(\lambda),N))$ which was shown to be an isomorphism. It follows from ([5], 1.16) that for each $V \in \mathrm{Ob}\, \underline{H}_{\hat{\lambda}}$ that we have an isomorphism

$(\delta V) \otimes_{U(\underline{g})} M(\lambda) \xrightarrow{\sim} \delta(V \otimes_{U(\underline{g})} M(\lambda))$.

Taking $V = L(M(\lambda),N)^\eta$ we have that

$\delta V = \delta(L(M(\lambda),N)^\eta) \cong (\delta L(M(\lambda),N))^\eta \cong L(M(\lambda),\delta N)^\eta$ and so

$\delta V \otimes_{U(\underline{g})} M(\lambda) = \varepsilon\delta N$ and the first isomorphism gives

$\varepsilon\delta N \xrightarrow{\sim} \delta\varepsilon N$, hence (i).

One has $L(M(\lambda),\delta M(w\lambda))^{\eta} \cong L(-w\lambda,-\lambda)^{\eta} \cong L(-\lambda,-w\lambda)$ $\cong L(-w^{-1}\lambda,-\lambda) \cong L(M(\lambda),\ \delta M(w^{-1}\lambda))$. Hence (ii). Then (iii) follows from (i), and (iv) from (ii), (iii) since $L(w\lambda)$ is the image of any non-zero map from $M(w\lambda) \to \delta M(w\lambda)$.

4.4 PROPOSITION. - Let C be a completion functor on $\underline{O}_{\Lambda}$. Then $C = C_M$ where $M = \varepsilon\delta\ C(\delta M(\lambda))$.

Chose $N, N' \in \mathrm{Ob}\ \underline{O}_{\Lambda}$. Given a map $g : N \to N'$ we must show that there exists an isomorphism $\varphi_N : CN \xrightarrow{\sim} C_M N$ such that the diagram

$$\begin{array}{ccc} CN & \xrightarrow[\sim]{\varphi_N} & C_M N \\ \downarrow{\scriptstyle C(g)} & & \downarrow{\scriptstyle C_M(g)} \\ CN' & \xrightarrow[\sim]{\varphi_{N'}} & C_M N' \end{array}$$

commutes.

Through 2.7 there exist E, F, E', F' and maps h, h', g', g'', such that the diagram

$$(*) \qquad \begin{array}{ccccccc} 0 \to & N & \to & E \otimes \delta M(\lambda) & \xrightarrow{h} & F \otimes \delta M(\lambda) \\ & \downarrow{\scriptstyle g} & & \downarrow{\scriptstyle g'} & & \downarrow{\scriptstyle g''} \\ 0 \to & N' & \to & E' \otimes \delta M(\lambda) & \xrightarrow{h'} & F' \otimes \delta M(\lambda) \end{array}$$

commutes and has exact rows.

Now consider the diagram

$$
\begin{array}{ccccc}
C(F^*\otimes E\otimes M) & \xrightarrow{C(1\otimes h)} & C(F^*\otimes F\otimes M) & \searrow & \\
\wr\downarrow \varphi_{E\otimes M,F^*} & & \wr\downarrow \varphi_{F\otimes M,F^*} & & f^*\otimes f \mapsto \langle f^*,f\rangle \\
F^*\otimes C(E\otimes M) & \xrightarrow{1\otimes C(h)} & F^*\otimes C(F\otimes M) & & C\;M \\
\wr\downarrow 1\otimes\varphi_{M,E} & & \wr\downarrow 1\otimes\varphi_{M,F} & & f^*\otimes f \mapsto \langle f^*,f\rangle \\
F^*\otimes E\otimes CM & \xrightarrow{1\otimes k} & F^*\otimes F\otimes CM & \nearrow &
\end{array}
$$

where $M=\delta M(\lambda)$ and $k \in \mathrm{Hom}\,(E\otimes CM, F\otimes CM)$ is defined to make the bottom square commute. Note that the top square commutes by (2a) and that the right-hand square commutes by (2b,c). Now by definition we have

$$h \in \mathrm{Hom}_{\underline{g}}(E\otimes M, F\otimes M) \cong \mathrm{Hom}_{\underline{g}}(F^*\otimes E, L(M,M)).$$

Now since $M=\delta M(\lambda)$ it follows from say ([5], 1.15) that the map $U(\underline{g}) \longrightarrow L(M,M)$ defined by the action of $U(\underline{g})$ on M is surjective. Let $\theta(h)$ denote the image of h in $\mathrm{Hom}_{\underline{g}}(F^*\otimes E, U(\underline{g})/\mathrm{Ann}\, M)$ which results. We apply (3) with E replaced by $\theta(h)(F^*\otimes E)$ (which as $\underline{g}$ is reductive we can consider as an ad $\underline{g}$ stable submodule of $U(\underline{g})$). Using (2) it then follows from the above diagram that k is just the map defined through $\langle k(e\otimes m'), f^*\rangle = (\theta(h)(f^*\otimes e))m'$, $\forall\, e \in E, f^*\in F^*$, $m'\in CM$. In particular k is determined by h.

The above considerations can be applied to the maps h', g', g'' which say determine k', ℓ', ℓ''. We thus obtain from (*), the commuting diagram with exact rows

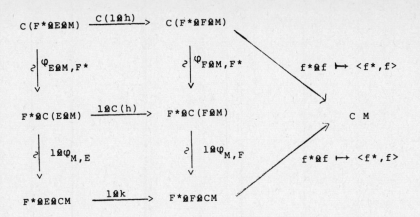

$$\begin{array}{ccccc} 0 \to CN & \longrightarrow & E \otimes C\delta M(\lambda) & \xrightarrow{k} & F \otimes C\delta M(\lambda) \\ \downarrow \ell & & \downarrow \ell' & & \downarrow \ell'' \\ 0 \to CN' & \longrightarrow & E' \otimes C\delta M(\lambda) & \xrightarrow{k'} & F' \otimes C\delta M(\lambda) \end{array}$$

where ℓ is equivalent to the map $C(g)$. Then $CN = \text{Ker } k$, $CN' = \text{Ker } k'$ and $\ell = \ell'|_{CN}$. We have thus shown that if C, C' are completion functors on $\underline{O}_\Lambda$ such that $C\delta M(\lambda) = C'\delta M(\lambda)$, then $C = C'$.

Finally take $C = C_M$. Then

$$C\delta M(\lambda) = L(M, \delta M(\lambda)) \otimes_{U(\underline{g})} M(\lambda)$$

$$\cong L(M(\lambda), \delta M)^\eta \otimes_{U(\underline{g})} M(\lambda)$$

$$= \varepsilon\delta(M), \text{ by definition of } \varepsilon.$$

Taking account of 4.3(i) the propositon follows.

4.5 COROLLARY. - <u>For all</u> $M, N, K \in \text{Ob } \underline{O}_\lambda$ <u>one has</u>

(i) $C_M C_N = C_L$ <u>where</u> $L = \varepsilon\delta(L(M, \varepsilon\delta(N)) \otimes_{U(\underline{g})} M(\lambda))$.

(ii) $L(M, L(N,K) \otimes_{U(\underline{g})} M(\lambda))$

$\cong L(\varepsilon\delta(L(M, \varepsilon\delta(N)) \otimes_{U(\underline{g})} M(\lambda)), K)$.

Indeed

$$C_M C_N \, \delta M(\lambda) = C_M(\varepsilon\delta(N))$$

$$= L(M, \varepsilon\delta(N)) \otimes_{U(\underline{g})} M(\lambda).$$

and (i) follows from 4.4. Then (ii) obtains from (i) by computing $C_M C_N K$.

4.6 From 4.5 (i) we see that the composition of completion functors is determined from a knowledge of the Harish-Chandra modules $L(M,N) : M,N \in \mathrm{Ob}\ \underline{O}_\Lambda$, whence the importance of these modules in the study of the $\underline{O}_\Lambda$ category. One may easily rederive 2.9 from 4.5 (i). Also 4.5 (ii) may be considered as expressing the associativity law $(C_M C_N)C_K = C_M(C_N C_K)$.

5. Kostant's problem.

5.1 Fix $w,y \in W_\lambda$ and set $K_{w,y} = L(L(w\lambda),L(y\lambda)) \otimes_{U(\underline{g})} M(\lambda)$, $K^\circ_{w,y} = \mathrm{Soc}\ K_{w,y}$. By ([12], 2.7, 2.18) we have that $K_{w,y} \neq 0$ implies that $d(K_{w,y}) = d(L(w\lambda)) = d(L(y\lambda))$ (where d denotes Gelfand-Kirillov dimension). Furthermore by ([12], 4.13) we have the rather deep result that $d(K_{w,y}/K^\circ_{w,y}) < d(L(w\lambda))$.

5.2 Let $\overset{\circ}{\Sigma}_\lambda$ denote the set of Duflo involutions of W_λ ([10], 3.4 and [12], Sect. 4). Let us choose $w,y \in W_\lambda$ such that $K_{w,y} \neq 0$. By ([5], 3.8) this exactly means in the language of ([12], 4.6) that w,y lie in the same right cell of W_λ. It is clear that there exist $x_i \in W_\lambda$ (not necessarily distinct) belonging to the same left cell as y and the same right cell as w such that

$$(*) \qquad K^\circ_{w,y} = \oplus L(x_i \lambda).$$

<u>Remark</u>. A formula for the x_i occuring in (*) is given in 5.11.

By definition of Σ°_λ there is exactly one element $\sigma \in \Sigma^\circ_\lambda$ belonging to the same left cell as w. Moreover by ([12], 3.5) we have that $K^\circ_{\sigma,w^{-1}} \cong L(w^{-1}\lambda)$.

THEOREM. <u>Choose</u> σ, w, $y \in W_\lambda$ <u>as above</u>. <u>Then</u>

$$L(L(w\lambda),L(y\lambda)) \cong \bigoplus L(L(\sigma\lambda),L(x_i\lambda))$$

<u>where the</u> x_i <u>are defined by</u> (*).

We apply 4.5 (ii) with $M = L(\sigma\lambda)$, $N = L(w\lambda)$, $K = L(y\lambda)$. Through the remarks in 5.1 we obtain that $L(L(\sigma\lambda),K_{w,y}) \xrightarrow{\sim} L(L(\sigma\lambda),K^\circ_{w,y}) \cong \oplus\, L(L(\sigma\lambda),L(x_i\lambda))$.

Again by 4.3 (iv) one has that

$$L(\varepsilon\delta(L(L(\sigma\lambda),\varepsilon\delta(L(w\lambda))) \otimes_{U(\underline{g})} M(\lambda)),L(y\lambda))$$

$$\cong L(\varepsilon\delta(K_{\sigma,w^{-1}}),L(y\lambda))$$

$$\cong L(\varepsilon\delta(K^\circ_{\sigma,w^{-1}}),L(y\lambda))$$

$$\cong L(L(w\lambda),L(y\lambda)),$$

as required.

5.3 Recall that $L(L(y\lambda),L(y\lambda))$ is a primitive Noetherian ring and so admits a ring of fractions.

By ([10], 3.4) the embedding $U(\underline{g})/J(y\lambda) \xhookrightarrow{i(y)} L(L(y\lambda),L(y\lambda))$ extends to an embedding Fract i(y) of rings of fractions and this is surjective if $y \in \Sigma^\circ_\lambda$.

COROLLARY - <u>Choose</u> $\sigma \in \Sigma^\circ_\lambda$ <u>belonging to the left cell of</u> W_λ <u>containing</u> y. <u>If</u> Fract i(y) <u>is surjective</u>, <u>then</u> $L(L(\sigma\lambda),L(\sigma\lambda)) \cong L(L(y\lambda),L(y\lambda))$.

Take w=y in 5.2. The hypothesis just says that $K^\circ_{y,y} = L(\sigma\lambda)$ and so the assertion follows from 5.2.

5.4 The above result extends non-trivially the set of all $y \in W_\lambda$ for which i(y) is surjective. (Kostant suggested that

$U(g)/\mathrm{Ann}\ M \to L(M,M)$ is surjective for any simple module M ; but this is almost never true). For example, one knows ([5], 4.4) that $i(y)$ is surjective when y takes the form $y=w_{B'}w_\lambda$ with $B' \subset B_\lambda$ (notation [5], 1.2). Thus the above result shows that $i(y)$ is surjective if Fract $i(y)$ is surjective and y belongs to a left cell containing an element of the form $w_{B'}w_\lambda$. Hopefully one could prove that $i(\sigma)$ is surjective for every $\sigma \in \Sigma^\circ_\lambda$ and more generally that $L(L(\sigma\lambda),L(w\lambda))$ is contained in the image of $L(M(\sigma\lambda),M(w\lambda))$ in $L(M(\sigma\lambda),L(w\lambda))$. Since by ([5], 3.4) $L(M(\sigma\lambda),M(w\lambda))$ identifies with a submodule of $U(\underline{g})/\mathrm{Ann}\ M(\lambda)$, this would identify $L(L(\sigma\lambda),L(w\lambda))$ as a subquotient of $U(\underline{g})/\mathrm{Ann}\ M(\lambda)$.

5.4 We extend the involution $w \mapsto w^{-1}$ of W_λ to an involution $*$ on $\mathbb{Q}W_\lambda$ through linearity. Let $a(w) \in \mathbb{Q}W_\lambda$ (notation [12], 4.6) represent $\left[L(w\lambda)\right]$ under the conventions of 2.2. Then $\{a(w) : w \in W_\lambda\}$ is a basis for $\mathbb{Q}W_\lambda$ and we remark that ([9], 3.3) implies that $a(w)^* = a(w^{-1})$. Now let J be a primitive ideal of $U(\underline{g})$. A basic problem is to compute the multiplicities in the Jordan-Holder series of $U(\underline{g})/J$ considered us a $U(\underline{g})$-$U(\underline{g})$ bimodule. One may assume without loss of generality that $J \supset \mathrm{Ann}\ M(\lambda)$ and then it is equivalent to compute the multiplicities $\left[M(\lambda)/JM(\lambda):L(w\lambda)\right]$ for each $w \in W_\lambda$. It is convenient to define

$$e_J := \sum_{w \in W_\lambda} \left[M(\lambda)/JM(\lambda) : L(w\lambda)\right] a(w)$$

and then our problem is to determine e_J as an element of the group algebra $\mathbb{Q}W_\lambda$. Let $a \mapsto {}^t a$ denote a Chevalley antiautomorphism (notation [5], 1.3) of $U(\underline{g})$. Then ${}^tJ = J$ and ${}^tV(-w\lambda,-\lambda) \cong V(-\lambda,-w\lambda) \cong V(-w^{-1}\lambda,-\lambda)$ ([4], Prop. 9) and so $e_J = e_J^*$. For each $B' \subset B_\lambda$ one has

LEMMA. - ([12], 4.8). _If_ $J=J(w_B, w_\lambda \lambda)$, _then_

$$e_J = (w_B, w_\lambda)^{-1} a(w_B, w_\lambda) = (\det w_{B'}) w_\lambda\, a(w_B, w_\lambda).$$

Remark. We take this opportunity to note that a factor of $\det w_{B'}$ was missing from the right hand side of the equation stated in [12], Lemma 4.8.

5.5 The above result has a number of features which are unfortunately not reproduced in general. For example, it is certainly false that $e_{J(w\lambda)} = w^{-1}a(w)$, $\forall\, w \in W_\lambda$. The reason for this is rather interesting and derives from the following result. Let $\overline{M(\mu)} : \mu \in \Lambda$ denote the unique maximal submodule of $M(\mu)$.

LEMMA. - _For each_ $w \in W_\lambda$ _one has_

$$J(w\lambda)M(\lambda) = C_{w^{-1}}(\overline{M(w\lambda)}).$$

Recall that the action of $U(\underline{g})$ on $M(w\lambda)$ defines a map $U(\underline{g}) \to L(M(w\lambda), M(w\lambda))$ which is surjective ([5], 3.4). It follows that $L(M(w\lambda), \overline{M(w\lambda)}) = \{a \in U(\underline{g})/\text{Ann}\, M(w\lambda) \mid aM(w\lambda) \subset \overline{M(w\lambda)}\}$ $= J(w\lambda)/\text{Ann}\, M(\lambda)$. Hence $J(w\lambda)M(\lambda) = L(M(w\lambda), \overline{M(w\lambda)}) \otimes_{U(\underline{g})} M(\lambda)$ $= C_{w^{-1}}(\overline{M(w\lambda)})$.

5.6 We should like to use 5.5 to compute $[M(\lambda)/J(w\lambda)M(\lambda)]$. If we use 2.9 then it is enough to compute $[C_\alpha M]$ from $[M]$ at each step. If M is α-cofree, then $[C_\alpha M] = s_\alpha [M]$ by 2.2 and (*) assuming that this holds at each step we obtain $[C_{w^{-1}}(\overline{M(w\lambda)})] = w^{-1}[\overline{M(w\lambda)}] = w^{-1}(w - a(w)) = 1 - w^{-1}a(w)$, and so $[M(\lambda)/J(w\lambda)M(\lambda)] = w^{-1}a(w)$. This is just 5.4 in the special case when $w=w_B, w_\lambda$. A more refined analysis (perhaps along the lines of ([6], Sect. 4)) is needed to handle the general case since our assumption (*) is obviously too strong. On the other hand we can derive from 5.5 an old result of Duflo ([4], Prop. 6) which determines when $[J(w\lambda)M(\lambda) : L(w'\lambda)] \neq 0$. For this call a subset $S \subset W_\lambda$ hereditary if $w \in S$ implies $y \in S$ for all $y \geq w$ (Bruhat order).

LEMMA. - <u>For each</u> $w \in W_\lambda$, $S_w := \{w' \in W_\lambda \mid [J(w\lambda)M(\lambda) : L(w'\lambda)] \neq 0\}$ <u>is the smallest hereditary subset of</u> W_λ <u>containing the</u> $\{s_\alpha : \alpha \in B_\lambda \text{ s.t. } w\alpha \in R^+\}$.

Take a reduced decomposition $s_1 s_2 \dots s_k$ of w, set $w_i = s_i s_{i-1} \dots s_1 : i=1,2,\dots,k$ and define $M_o = \overline{M(w\lambda)}$, $M_i := C_{w_i}(\overline{M(w\lambda)})$, $S_i := \{w' \in W_\lambda \mid [M_i : L(w'\lambda)] \neq 0\}$. Choose $y \in W_\lambda$, $\alpha \in B_\lambda$. If $s_\alpha y > y$, then $C_\alpha L(y\lambda) = 0$ ([13], 3.5).

Otherwise by ([13], 3.5) we have that $[C_\alpha L(y\lambda) : L(w'\lambda)] \neq 0$ implies either $w' \geq y$ or $w' = s_\alpha y$. Again if M is α-free then $D_\alpha^+ M$ embeds in M, so by ([13], 3.4) we can compute <u>JH</u> $(C_\alpha M)$ by assuming that C_α is exact and computing $[C_\alpha L]$ for each simple subquotient L of M.

By say ([3], 7.5.23) one has $S_o = \{y \in W_\lambda \mid y \geq w\}$ and so is hereditary. Assume that we have shown that S_i is hereditary. By the above observations it follows that

$$S_{i+1} = S_i \cup \{s_{i+1} y \mid y \in S_i \,,\, s_{i+1} y < y\}.$$

Observe that either $s_{i+1} y < y$ or $z := s_{i+1} y > y$ and then $y \in S_i \Rightarrow z \in S_i$ and $y = s_{i+1} z$. It follows that S_{i+1} is hereditary and is generated as a hereditary subset by $s_{i+1} S_i$. Consequently $S_w := S_k$ is the smallest hereditary subset of W_λ containing $\{w^{-1}y \mid y \in W_\lambda,\ y > w\}$. In this we can assume without less of generality that $\ell(y) = \ell(w)+1$ and so we can write $y = ws_\gamma$ with $\gamma \in R^+$ such that $w\gamma \in R^+$. Write

$$\gamma = \sum_{\alpha \in B_\lambda} k_\alpha \alpha \ :\ k_\alpha \in \mathbb{N}$$

If $k_\alpha \neq 0$, then $s_\alpha \leq s_\gamma$ and moreover one can obviously find $\alpha \in B_\lambda$ with $k_\alpha \neq 0$ and $w\alpha \in R^+$. In view of our previous observation, this proves the lemma.

Remarks. Take $\alpha \in B_\lambda$, $w \in W_\lambda$. If $s_\alpha w > w$, then $\overline{M(w\lambda)} \supset C_\alpha(\overline{M(s_\alpha w\lambda)})$ (cf. [13], 5.1) and so $C_{w^{-1}}(\overline{M(w\lambda)}) \supset C_{y^{-1}}(\overline{M(y\lambda)})$ where $y=s_\alpha w$. Consequently $J(s_\alpha w\lambda) \subset J(w\lambda)$ and by induction $J(w'\lambda) \subset J(w''\lambda)$ wherever $S(w') \supset S(w'')$ (notation [7], 2.1). This is an old result of Duflo ([4], Sect. 3, Cor. 1). Actually one might have used the more precise and deep result ([13], 5.1) that $\overline{M(w\lambda)} = C_\alpha(\overline{M(s_\alpha w\lambda)}) + M(s_\alpha w\lambda)$; but we have not been able to do so. Now if $w\alpha \in R^+$, then $S(w_\lambda s_\alpha) \supset S(w)$ and so $J(w_\lambda s_\alpha\lambda) \subset J(w\lambda)$. Conversely if $J(w_\lambda s_\alpha\lambda) \subset J(w\lambda)$, then $0 \neq [J(w_\lambda s_\alpha\lambda)M(\lambda) : L(s_\alpha\lambda)] \leq [J(w\lambda)M(\lambda) : L(s_\alpha\lambda)]$ and so $w\alpha \in R^+$ by 5.5. This is how Duflo obtained the Borho-Janzten "τ-invariant", and it is essentially the best information one can obtain from just $\underline{JH}\,(J(w\lambda)M(\lambda))$.

5.7 As an alternative to 5.5, we remark that a similar argument gives the

LEMMA.- _For each_ $w \in W_\lambda$,

$M(\lambda)/J(w\lambda)M(\lambda)$ _is the unique submodule of_ $C_{w^{-1}}L(w\lambda)$ _admitting_ $L(\lambda)$ _as a subquotient_.

5.8 Using 2.9 one may analyse how $M(\lambda)/J(w\lambda)M(\lambda)$ is "build up" from the action of the completion functors C_α on

$L(w\lambda)$. In particular using 2.2 one may easily establish Vogan's result ([14], Sect. 3) describing the ordering of the primitive ideals, namely

LEMMA. - (notation [11], Sect. 5). <u>For each</u> $w,w' \in W_\lambda$ <u>one has</u> $J(w'\lambda) \supset J(w\lambda)$ <u>if and only if</u> $a(w') \in [W_\lambda a(w)]$.

5.9 By ([9], 4.5) the $\mathbb{Q}[W_\lambda a(w)]$: $w \in W_\lambda$, are left ideals of $\mathbb{Q}W_\lambda$ which by 5.8 determine the ordering of the primitive ideals. Since $\mathbb{Q}W_\lambda$ is a semi-simple Artinian ring, each left ideal of $\mathbb{Q}W_\lambda$ takes the form $(\mathbb{Q}W_\lambda)e$ with $e^2=e$ and $e^*=e$ (noting that * is just the adjoint with respect to the W_λ invariant bilinear form on $\mathbb{Q}W_\lambda$ defined through $(y,z) \mapsto \operatorname{tr} y^{-1}z$, $\forall\, y,z \in W_\lambda$), moreover such an e is uniquely determined. In particular we can write $\mathbb{Q}[W_\lambda a(w)] = \mathbb{Q}W_\lambda e_w$ for some uniquely determined self-adjoint idempotent e_w which by 5.8 determines the ordering of the primitive ideals. On the other hand it is clear that the $e_{J(w\lambda)}$ which are self-adjoint (5.4) also determine the ordering of primitive ideals. It is therefore natural to guess that $e_{J(w\lambda)} = n_w e_w$ for some rational number n_w. If so this would completely determine the multiplicities of simple factors in $M(\lambda)/J(w\lambda)M(\lambda)$ (or equivalently in $U(\mathfrak{g})/J(w\lambda)$). When $w=w_B, w_\lambda$, this conjecture does hold and we even have $n_w = \operatorname{card} W_B$, in this case. This assertion follows from 5.4 and ([9], 5.11, 5.12 where one must note that $-\lambda$ is dominant). We have checked this conjecture up to type A_3 and we remark that it implies $(\underline{C}_1)$, $(\underline{C}_2)$ below. One can show (see below) that n_w is an integer > 0 and from 5.8 that

$$e_{J(w\lambda)} = e_w e_{J(w\lambda)} = e_{J(w\lambda)} e_w .$$

5.10 Recall that $\{a(w) \mid w \in W_\lambda\}$ is a basis for W_λ and let us introduce a vector space gradation of $\mathbb{Q}W_\lambda$ by defining

$a(w)$ to have degree $d(w) := \text{card } R^+ - d(L(w\lambda))$. We remark that through ([11], 5.1) $d(w)$ is determined by $a(w)$. Moreover by ([9], 3.1) $d(w) = d(w^{-1})$ and by ([11], 5.1) $a(w)a(w')$ has degree $\geq \max\{d(w), d(w')\}$. We write

$$a(w)a(w') \doteq \sum_{y \in W_\lambda} c_y a(y) \ : \ c_y \in \mathbb{Q},$$

if $a(w)a(w') - \sum c_y a(y)$ is of gradation $> \min\{d(w), d(w')\}$, and we write $a(y) \dot{\in} [a(w)a(w')]$ if $c_y \neq 0$. Note that

$$a(w)a(w') \doteq 0 \quad \text{if} \quad d(w) \neq d(w').$$

Given $y \in W_\lambda$, let $C(y)$ (resp. $C'(y)$) denote the left (resp. right) cell containing y (c.f. [12], 4.5).

LEMMA. - Suppose $a(x) \dot{\in} [a(w^{-1})a(y)]$. Then $x \in C(y)$ and $x^{-1} \in C(w)$.

Obviously $a(w^{-1})a(y) \in [W_\lambda a(y)] \cap [a(w^{-1})W_\lambda]$. So if $a(x) \dot{\in} [a(w^{-1})a(y)]$ one has $x \in C(y) \cap C'(w^{-1})$. Finally $x \in C'(w^{-1}) \Leftrightarrow x^{-1} \in C(w)$.

5.11 Take $\sigma \in \Sigma_\lambda^\circ$ (notation [10], 3.4). Then by definition of Σ_λ° there is a uniquely determined primitive ideal J (which we can write as $J(\sigma\lambda)$) such that $M(\lambda)/JM(\lambda) \doteq L(\sigma\lambda)$ (notation [12], 2.11). In particular $L(\sigma\lambda)$ is the unique subquotient of $M(\lambda)/JM(\lambda)$ having the maximal Gelfand-Kirillov dimension. It follows from the remarks in 5.10 that if $e_J^2 = n_J e_J : n_J \in \mathbb{Q}$ then $a(\sigma)^2 = n_J a(\sigma)$. Moreover since

$$a(\sigma) = \sigma + \sum_{w > \sigma} a(\sigma, w) w \ : \ a(\sigma, w) \in \mathbb{Z}$$

it follows that $n_J \in \mathbb{N}^+$. Quite independent of this, it follows from 5.10 that if $a(x) \dot{\in} [a(\sigma)^2]$ then x belongs to the intersection of the left and right cells of W_λ containing σ. This is essentially a small set, often reduced to $\{\sigma\}$ - for example in type A_n. We thus formulate the conjecture.

($\underline{C}_1$) <u>For each</u> $\sigma \in \Sigma^\circ_\lambda$ <u>one has</u> $a(\sigma)^2 \doteq n_\sigma \, a(\sigma)$ <u>with</u> $n_\sigma \in \mathbb{N}^+$.

Recalling ([12], 4.10) and that the Janzten conjecture has now been claimed to have been proved, it follows that Σ°_λ is determined by the Kazhdan-Lusztig polynomials (by the JKL data in the language of [12]). Since the $a(\sigma)$ are also determined by the Kazhdan-Lusztig polynomials evaluated at $q=1$ (by the Jantzen matrix in the language of [12]) we see that ($\underline{C}_1$) can be regarded as a purely combinatorial problem, though this not the way it should be solved! Again Lusztig informed me that (at least in type A_n) he has considered the relation $a(\sigma)^2 \doteq n_\sigma \, a(\sigma)$; but now in the Hecke algebra for which n_σ becomes a polynomial in q, and has conjectured a geometric interpretation for these polynomials.

Let us recall an earlier conjecture. ([12], ($\underline{C}_3$)) which again reduces to a purely combinatorial question ; namely

($\underline{C}_2$) <u>For each</u> $\sigma \in \Sigma^\circ_\lambda$, $a(\sigma)$ <u>is a cyclic vector in its left (or right) cell.</u>

The significance of ($\underline{C}_1$) and ($\underline{C}_2$) is that they can be used to give the following rather nice interpretation of ([12], 4.14). For this take $\theta_\alpha : \alpha \in B_\lambda$ to be the functor of coherent continuation across the α-wall (see [6], 3.2 for example). Then in the conventions of 2.2 we have $[\theta_\alpha M] = [M](1+s_\alpha)$, $\forall\, M \in \mathrm{Ob}\, \underline{O}_\lambda$. Applied to ([12], 3.9 (ii)) we obtain as in ([12], 4.14) using ([12], 4.13) and 5.10 the

THEOREM. - <u>Fix</u> $\sigma \in \Sigma_\lambda^\circ$ <u>and suppose</u> $(\underline{C}_1)$, $(\underline{C}_2)$ <u>hold. Choose</u> $w \in C(\sigma)$ <u>and</u> $y \in C'(w)$. <u>Then</u>

$$\mathrm{Soc}\ \mathcal{L}(L(w\lambda),L(y\lambda)) = \oplus L(x_i\lambda),$$

<u>where the</u> x_i <u>occuring in the above sum are determined through the formula</u>

$$(*) \qquad a(w^{-1})a(y) \doteq n_\sigma(\Sigma\, a(x_i)).$$

<u>Remarks</u>. Combined with ([12], 3.5 (i)) it follows taking $w=\sigma$ in (*) that $a(\sigma)a(y) \doteq n_\sigma\ a(y)$ and this shows exactly what element (namely $a(y)$) of $\mathbb{Q}W_\lambda$ must be used to generate $a(y)$ from the cyclic vector $a(\sigma)$. Again take $y=w$ in (*). From the embedding $U(\mathfrak{g})/J(\sigma\lambda) \hookrightarrow \mathcal{L}(L(w\lambda),L(w\lambda))$ it follows that at least one (and exactly one) x_i equals σ. This means that we can compute Σ_λ° from the just the Jantzen matrix (instead of the more refined JKL data). Indeed one simply determines all $w \in W_\lambda$ such that $a(w^{-1})a(w) \doteq a(x)$ (i.e. where the right hand side reduces to exactly one term) and then the set of x so obtained is precisely Σ_λ°.

If $y \notin C'(w)$, then $\mathcal{L}(L(w\lambda),L(y\lambda)) = 0$; but one need not have that $a(w^{-1})a(y) \doteq 0$. Again even if $y \in C'(w)$, the higher order terms in the product $a(w^{-1})a(y)$ do not necessarily give the terms of smaller Gelfand-Kirillov dimension in $\mathcal{L}(L(w\lambda),L(y\lambda))$. On the other hand one can define a multiplication * in $\mathbb{Q}W_\lambda$ through

$$a(w^{-1}) * a(y) = \sum_{x \in W_\lambda} m_x a(x),\ \forall\ y,w \in W_\lambda,$$

where

$$m_x := \left[\mathcal{L}(L(w\lambda),L(y\lambda)) \otimes_{U(\mathfrak{g})} M(\lambda) : L(x\lambda)\right].$$

Unfortunately this multiplication does not appear to have any nice properties, for example it is not associative now can it be defined in W_λ. Yet this multiplication is associative if one retains only terms a(x) satisfying $d(x) \leq \min\{d(w),d(y)\}$. As in 5.2 this follows from 4.5 (i) and the associativity of multiplication of completion functors, or of course also from 5.11 (given that $(\underline{C}_1)$, $(\underline{C}_2)$ hold).

Index of notation

Symbols appearing frequently throughout the text are given below in order of appearance. (See also [5], [13]).

1.1. $\underline{O}$.

1.2. $\underline{g}$, $\underline{h}$, R, B, P(R), $\underline{h}^*$, λ, Λ, $\underline{O}_\Lambda$, R_λ, W_λ.

1.3. $L(\mu)$, $J(\mu)$.

1.4. $L(M,N)$, $M(\lambda)$.

2.1. B_λ, R_λ^+, C_α, D_α^-, $X_{-\alpha}$, δ, D_α^+.

2.2. [M], $\underline{O}_{\hat{\mu}}$.

2.3. $T_\alpha^{M,N}$.

2.4. η.

2.5. $L(-\lambda,-\mu)$.

2.6. T_α^M.

2.7. $\mathcal{E}$.

3.1. κ_α^M.

3.2. D_α.

4.3. ε, $\underline{H}_{\hat{\lambda}}$.

5.1. $K_{w,y}$, $K_{w,y}^o$.

5.4. $a(w)$, ${}^t a$.

5.10. $d(w)$.

REFERENCES

[1] A. Bouaziz, Sur les représentations des algèbres de Lie semi-simples construites par T. Enright, pp. 57-68, LN 880, Springer-Verlag, Berlin |Heidelberg| New York, 1981.

[2]. V.V. Deodhar, On a construction of representations and a problem of Enright, Invent. Math., 57 (1980) 101-118.

[3]. J. Dixmier, Algèbres enveloppantes, Cahiers Scientifiques, XXXVII, Gauthier-Villars, Paris, 1974.

[4]. M. Duflo, Sur la classification des idéaux primitifs dans l'algèbre enveloppante d'une algèbre de Lie semi-simple, Ann. of Math. 105 (1977) 107-130.

[5]. O. Gabber and A. Joseph, The Bernstein-Gelfand-Gelfand resolution and the Duflo sum formula, Compos. Math., 43 (1981) 107-131.

[6]. O. Gabber and A. Joseph, Towards the Kazhdan-Lusztig conjecture, Ann. Ec. Norm. Sup., 14 (1981) 261-302.

[7]. A. Joseph, On the annihilators of the simple subquotients of the principle series, Ann. Ec. Norm. Sup. 10 (1977) 419-440.

[8]. A. Joseph, Gelfand-Kirillov dimension for the annihilators of simple quotients of Verma modules, J. Lond. Math. Soc., 18 (1978) 50-60.

[9]. A. Joseph, W-module structure in the primitive spectrum of the enveloping algebra of a semi-simple Lie algebra, pp. 116-135, LN 728, Springer-Verlag, Berlin |Heidelberg| New York, 1979.

[10]. A. Joseph, Goldie rank in the enveloping algebra of a semi-simple Lie algebra, I, J. Alg., 65 (1980) 269-283.

[11]. A. Joseph, Goldie rank in the enveloping algebra of a semi-simple Lie algebra, II, J. Alg. 65 (1980) 284-306.

[12]. A. Joseph, Goldie rank in the enveloping algebra of a semi-simple Lie algebra, III, J. Alg. 73 (1981) 295-326.

[13]. A. Joseph, The Enright functor in the Bernstein-Gelfand-Gelfand category O, Invent. Math., (to appear).

[14]. D.A. Vogan, Ordering of the primitive spectrum of a semi-simple Lie algebra, Math. Ann. 248 (1980) 195-203.

Department of Theorical Mathematics
The Weizmann Institute of Science
Rehovot 76100, Israël

and

Laboratoire de Mathématiques fondamentales
(Equipe de Recherche associée au CNRS)
Université de Pierre et Marie Curie
France

MINIMAL K-TYPE FORMULA

A. W. Knapp*

In an effort to attach new invariants to group representations, D. A. Vogan introduced in [9] a notion of minimal or "lowest" K-types for representations of semisimple Lie groups and used it as a starting point for several deep investigations in representation theory. What we shall do here is to announce a simple formula for all the minimal K-types of the standard representations induced from parabolic subgroups MAN when the inducing data include a discrete series or nondegenerate limit of discrete series representation of M and when the total group is linear. If we anticipate that certain results of Vogan's extend to all of our representations, then it follows from Theorem 5 of [5] that we obtain a minimal K-type formula for all irreducible admissible representations of linear semisimple groups in terms of their Langlands parameters [7].

Some applications of our formula appear in the joint paper [4] with B. Speh.

Let G be a linear connected semisimple Lie group, let K be a maximal compact subgroup, and let $\mathfrak{g}$ and $\mathfrak{k}$ be the Lie algebras of G and K. Fix a maximal abelian subspace $\mathfrak{b}$ of $\mathfrak{k}$, and let

$$\Delta_K = \{\text{roots of } (\mathfrak{k}^{\mathbb{C}}, \mathfrak{b}^{\mathbb{C}})\} \subseteq (i\mathfrak{b})'$$

$$\Delta_K^+ = \text{some positive root system for } \Delta_K$$

$$\rho_K = \text{half the sum of the members of } \Delta_K^+.$$

---

* Research supported by grants from the National Science Foundation through the University of Chicago and Cornell University.

To each dominant integral form Λ on $\mathfrak{b}^{\mathbb{C}}$, we associate the irreducible representation τ_Λ of K with highest weight Λ. We introduce an inner product $\langle\cdot,\cdot\rangle$ and a norm $|\cdot|$ on $(i\mathfrak{b})'$ in the usual way.

If π is an irreducible admissible representation of G, we say that τ_Λ is a minimal K-type of π if, among all irreducible representations $\tau_{\Lambda'}$ occurring in $\pi|_K$, $|\Lambda' + 2\rho_K|^2$ is minimized for $\Lambda' = \Lambda$. Existence of minimal K-types for any π is clear; there may be several. It is important to note that this notion is independent of the choice of the positive system Δ_K^+.

Let $P = MAN$ be the Langlands decomposition of a cuspidal parabolic subgroup of G, let σ be a discrete series or nondegenerate limit of discrete series[1] representation of M, let ν be a complex-valued linear functional on the Lie algebra of A, and form the induced representation[2]

$$U(P,\sigma,\nu) = \mathrm{ind}_{MAN}^{G}(\sigma \otimes e^{\nu} \otimes 1) . \tag{0.1}$$

The minimal K-types of $U(P,\sigma,\nu)$ are independent of ν, and we shall give a formula for them. For the precise formula we need to define suitably compatible orderings for various root systems that occur. But if we ignore this difficulty for the moment, we can give the formula approximately. Disregarding the possible disconnectedness of M, let λ be the Blattner parameter of σ; this is the highest weight of the minimal $K \cap M$ type of σ. Then the minimal K-types τ_Λ of $U(P,\sigma,\nu)$ are given by

---

[1] See §§1 and 12 of [6] for the definition and elementary properties of limits of discrete series and nondegeneracy.

[2] The notation refers to unitary induction with G operating on the left.

$$\Lambda = \lambda - E(2\rho_K) + 2\rho_{K_r} + \mu . \qquad (0.2)$$

Here E is the orthogonal projection to the subspace orthogonal to the Lie algebra of M. The term $2\rho_{K_r}$ refers to the $2\rho_K$ for a certain split subgroup G_r of G determined by G and M, and μ refers to any of several fine[3] K_r-types for G_r. In practice the group G_r is often locally just a product of copies of $SL(2,\mathbb{R})$, and μ is easy to understand; in principle G can be split and P can be a minimal parabolic subgroup, in which case $G_r = G$, $\Lambda = \mu$, and the formula gives no information.

The notation needed to make sense out of (0.2) and to define the compatible orderings is assembled in §1. The reader is asked to think first in terms of the case that $\operatorname{rank} G = \operatorname{rank} K$, where $\Delta_K \subseteq \Delta$ and where the other notation simplifies greatly. The precisely stated minimal K-type formula appears as Theorems 1 and 2 in §2, and Theorem 4 of §2 gives additional information about μ when G_r is locally a product of copies of $SL(2,\mathbb{R})$.

1. Notation

We continue with G, $\mathfrak{g}$, K, $\mathfrak{k}$, $\mathfrak{b}$, and Δ_K as in the introduction, but we postpone defining the positive system Δ_K^+. Let

θ = Cartan involution of $\mathfrak{g}$ determined by $\mathfrak{k}$

$\mathfrak{g} = \mathfrak{k} \oplus \mathfrak{p}$: corresponding Cartan decomposition

$\mathfrak{t}$ = centralizer of $\mathfrak{b}$ in $\mathfrak{g}$.

---

[3] See §1 below for a definition of "fine." The notion was introduced by Bernstein, Gelfand, and Gelfand [1] and developed further by Vogan [10]. We use some of Vogan's results.

Here $\mathfrak{t}$ is a maximally compact θ-stable Cartan subalgebra of $\mathfrak{g}$ and is of the form $\mathfrak{t} = \mathfrak{b} \oplus \mathfrak{a}$, where $\mathfrak{a} = \mathfrak{t} \cap \mathfrak{p}$. (This $\mathfrak{a}$ will usually not coincide with the Lie algebra of the group A in the introduction.) Let

$$B = \exp \mathfrak{b}$$

$$\Delta = \{\text{roots of } (\mathfrak{g}^{\mathbb{C}}, \mathfrak{t}^{\mathbb{C}})\}$$

$$\Delta_B = \{\text{roots of } (\mathfrak{g}^{\mathbb{C}}, \mathfrak{t}^{\mathbb{C}}) \text{ vanishing on } \mathfrak{a}\}.$$

The root vectors for the members of Δ_B lie either in $\mathfrak{k}^{\mathbb{C}}$ or in $\mathfrak{p}^{\mathbb{C}}$, and we call the corresponding roots <u>compact</u> or <u>noncompact</u>, accordingly. Let

$$\Delta_{B,c} = \{\text{compact roots in } \Delta_B\}$$

$$\Delta_{B,n} = \{\text{noncompact roots in } \Delta_B\}.$$

One can show that restriction from $\mathfrak{t}^{\mathbb{C}}$ to $\mathfrak{b}^{\mathbb{C}}$ carries $\Delta - \Delta_{B,n}$ onto Δ_K; consequently we can regard $\Delta_{B,c}$ as a subset of Δ_K.

To characterize the M of our parabolic subgroup up to conjugacy, it is enough (by Harish-Chandra's construction in [2]) to specify a conjugacy class of θ-stable Cartan subalgebras in $\mathfrak{g}$, and this conjugacy class in turn is determined by specifying a sequence $\alpha_1, \ldots, \alpha_\ell$ of strongly orthogonal members of $\Delta_{B,n}$. (See §2 of Schmid [8] for an exposition.) Thinking of the effect of a Cayley transform, we say that

$$\text{a root in } \Delta \text{ is } \begin{cases} \underline{\text{real}} & \text{if in } \sum \mathbb{R}\alpha_j \oplus \mathfrak{a}' \\ \underline{\text{imaginary}} & \text{if orthogonal to } \sum \mathbb{R}\alpha_j \oplus \mathfrak{a}' \\ \underline{\text{complex}} & \text{otherwise.} \end{cases}$$

Let

$\Delta_r = \{\text{real roots in } \Delta\}$

$\mathfrak{b}_r = \sum \mathbb{R} i H_{\alpha_j}$, where H_{α_j} is dual in $\mathfrak{t}^{\mathbb{C}}$ to α_j

$\mathfrak{b}_- =$ orthocomplement in $\mathfrak{b}$ to $\mathfrak{b}_r$, so that $\mathfrak{b} = \mathfrak{b}_- \oplus \mathfrak{b}_r$

$\mathfrak{t}_r = \mathfrak{b}_r \oplus \mathfrak{a}$

$E =$ orthogonal projection of $(\mathfrak{t}^{\mathbb{C}})'$ onto $(\mathfrak{t}_r^{\mathbb{C}})'$.

The subalgebra

$$\mathfrak{g}_r = \mathfrak{g} \cap (\mathfrak{t}_r^{\mathbb{C}} \oplus \sum_{\beta \in \Delta_r} \mathbb{C} X_\beta)$$

is a θ-stable reductive subalgebra of $\mathfrak{g}$ that is split over $\mathbb{R}$.

Let G_r be the analytic subgroup of G with Lie algebra $\mathfrak{g}_r$. The group $K_r = K \cap G_r$ is a maximal compact subgroup of G_r, and its Lie algebra is $\mathfrak{k}_r = \mathfrak{k} \cap \mathfrak{g}_r$. Moreover, $\mathfrak{b}_r$ is a maximal abelian subspace of $\mathfrak{k}_r$, $\mathfrak{t}_r$ is a maximally compact θ-stable Cartan subalgebra of $\mathfrak{g}_r$, and Δ_r is the root system of $(\mathfrak{g}_r^{\mathbb{C}}, \mathfrak{t}_r^{\mathbb{C}})$.

To obtain M, we build a Cayley transform $\underset{\sim}{c}$ out of the roots $\alpha_1, \ldots, \alpha_\ell$ and construct a new θ-stable Cartan subalgebra $\mathfrak{g} \cap \underset{\sim}{c}(\mathfrak{t}^{\mathbb{C}})$, as in [8]. Then we construct M and its Lie algebra $\mathfrak{m}$ in the standard way [2]. With

$$\Delta_- = \{\beta \in \Delta \mid \beta|_{\mathfrak{t}_r} = 0\} ,$$

$\mathfrak{m}$ is equal to the intersection of $\mathfrak{g}$ with

$$\mathfrak{m}^{\mathbb{C}} = \mathfrak{b}_-^{\mathbb{C}} \oplus \sum_{\beta \in \Delta_-} \mathbb{C}\, \underset{\sim}{c}(X_\beta) .$$

Each root vector $\underset{\sim}{c}(X_\beta)$ for $\mathfrak{m}^{\mathbb{C}}$ is either in $\mathfrak{k}^{\mathbb{C}}$ or in $\mathfrak{p}^{\mathbb{C}}$, and we call β <u>M-compact</u> or <u>M-noncompact</u> accordingly. Let

$$\Delta_{-,c} = \{\text{M-compact roots in } \Delta_-\}$$

$$\Delta_{-,n} = \{\text{M-noncompact roots in } \Delta_-\} .$$

Since $\mathfrak{t}_r \supseteq \mathfrak{a}$, we have $\Delta_- \subseteq \Delta_B$. However, $\Delta_{-,c}$ need not be contained in $\Delta_{B,c}$, since $\underset{\sim}{c}$ may move X_β from $\mathfrak{p}^{\mathbb{C}}$ to $\mathfrak{k}^{\mathbb{C}}$.

Every discrete series or limit of discrete series representation of M is known to be induced[4] from the subgroup

$$M^{\#} = M_e Z_M ,$$

where M_e is the identity component of M and Z_M is the center of M. The algebra $\mathfrak{b}_-$ is a compact Cartan subalgebra of $\mathfrak{m}$; let $B_- = \exp \mathfrak{b}_-$. By Lemma 2.1c of [6], we have

$$M^{\#} = M_e M_r , \tag{1.1}$$

where M_r is defined as the finite abelian group

$$\begin{aligned} M_r = F(B_-) &= \operatorname{span}\{\gamma_{\underset{\sim}{c}(\beta)} \mid \beta \in \Delta \text{ and } \beta|_{\mathfrak{b}_-} = 0\} \\ &= \operatorname{span}\{\gamma_{\underset{\sim}{c}(\beta)} \mid \beta \in \Delta_r\} . \end{aligned}$$

The element $\gamma_{\underset{\sim}{c}(\beta)}$ is the element of G corresponding to the matrix $\begin{pmatrix} -1 & 0 \\ 0 & -1 \end{pmatrix}$ in the $SL(2,\mathbb{R})$ subgroup built from the root $\underset{\sim}{c}(\beta)$. The group M_r is the M of a minimal parabolic subgroup of the split group G_r.

Let $\sigma^{\#}$ be a discrete series or nondegenerate limit of discrete series representation of $M^{\#}$. Because of (1.1), it follows from §1 of [6] that $\sigma^{\#}$ is determined by a triple (λ_0, C, χ), where

λ_0 is a Harish-Chandra parameter of $\sigma^{\#}$ relative to $(\mathfrak{m}, \mathfrak{b}_-)$

C is a Weyl chamber with respect to which λ_0 is dominant

χ is the scalar $\sigma|_{M_r}$.

---

4 See §1 of [6] for an exposition in the discrete-series case and a proof in the limits-of-discrete-series case.

This triple will allow us to define compatible positive systems for the various root systems we have introduced.

Define $(\Delta_-)^+$ so that C is the dominant chamber, and define $\rho_{-,c}$ and $\rho_{-,n}$ as the corresponding half sums of positive members of $\Delta_{-,c}$ and $\Delta_{-,n}$. The Blattner parameter of $\sigma^\#$, given by

$$\lambda = \lambda_0 + \rho_{-,n} - \rho_{-,c} ,$$

has the property that the unique minimal $K \cap M^\#$ type of $\sigma^\#$ is

$$\sigma_\lambda = \begin{cases} \text{irreducible representation with} & \\ \quad \text{highest weight } \lambda & \text{on } K \cap M_e \\ \chi & \text{on } M_r . \end{cases}$$

(This follows from Theorem 1.3 of Hecht and Schmid [3].)

To define Δ^+, let

$$\Delta_1^+ = \{\beta \in \Delta \mid \langle \lambda_0, \beta \rangle > 0\}$$

$$\Delta_0 = \{\beta \in \Delta \mid \langle \lambda_0, \beta \rangle = 0\}$$

$$\mathfrak{t}_0 = \Big(\sum_{\beta \in \Delta_0} \mathbb{C} H_\beta \Big) \cap \mathfrak{g}$$

$$\Delta_{-,0} = \Delta_- \cap \Delta_0 .$$

Then $\Delta_{-,0}$ is generated by $(\Delta_-)^+$ simple roots $\epsilon_1, \ldots, \epsilon_k$. Since $\mathfrak{t}_0 \supseteq \mathfrak{b}_r \oplus \mathfrak{a}$, we have

$$\mathfrak{t}_0 = (\mathfrak{t}_0 \cap \mathfrak{b}_-) \oplus \mathfrak{b}_r \oplus \mathfrak{a} .$$

Therefore the following list provides an ordered basis of real elements in $(\mathfrak{t}_0^{\mathbb{C}})'$:

$$\epsilon_1, \ldots, \epsilon_k, \text{ orthogonal basis of remainder of } i(\mathfrak{t}_0 \cap \mathfrak{b}_-)',$$

$$\alpha_1, \ldots, \alpha_\ell, \text{ basis of } \mathfrak{a} . \tag{1.2}$$

We use this ordered basis to define a lexicographic ordering. This ordering defines Δ_0^+, and we take

$$\Delta^+ = \Delta_1^+ \cup \Delta_0^+ .$$

Then one can check that Δ^+ is a positive system with $(\Delta_-)^+ \subseteq \Delta^+$ and that

$$\Delta_K^+ = \{\delta \in \Delta_K \mid \delta = \text{restriction to } \mathfrak{b}^{\mathbb{C}} \text{ of a member of } \Delta^+ - \Delta_{B,n}\}$$

is a positive system for Δ_K.

Finally the inclusions $\Delta_r \subseteq \Delta$ and $\Delta_{K_r} \subseteq \Delta_K$ define Δ_r^+ and $\Delta_{K_r}^+$ for us, and these definitions are compatible within G_r with the above construction for passing from G_r to K_r.

If $\sigma = \operatorname{ind}_{M^\#}^{M}(\sigma^\#)$, then the representation (0.1) satisfies

$$U(P,\sigma,\nu)|_K = \operatorname{ind}_{K\cap M^\#}^{K}(\sigma^\#) ,$$

and we shall work with it in this form. Correspondingly the restriction to K_r of the nonunitary principal series of G_r induced from data including a character ω of M_r is

$$\operatorname{ind}_{M_r}^{K}(\omega) .$$

A minimal K_r-type τ_μ in this case is called a <u>fine</u> K_r-type; τ_μ contains no other characters of M_r besides ω and its conjugates by the Weyl group.

2. Results

Now we come to the theorems. Let $\sigma^\# \leftrightarrow (\lambda_0, C, \chi)$ be a discrete series or nondegenerate limit of discrete series representation of $M^\#$, and let the notation and orderings be as in §1.

Theorem 1. Every minimal K-type τ_Λ of $\mathrm{ind}_{K\cap M^\#}^{K}(\sigma^\#)$ has Λ of the form

$$\boxed{\Lambda = \lambda - E(2\rho_K) + 2\rho_{K_r} + \mu} , \tag{2.1}$$

where τ_μ is a fine K_r-type whose restriction to M_r contains the character

$$\omega = \chi \cdot \exp(E(2\rho_K) - 2\rho_{K_r})|_{M_r} ; \tag{2.2}$$

here $\exp(E(2\rho_K) - 2\rho_{K_r})$ is a well-defined one-dimensional representation of $K_r \supseteq M_r$. Conversely every fine K_r-type τ_μ with $\tau_\mu|_{M_r} \supseteq \omega$ is such that Λ in (2.1) is integral; if Λ is also Δ_K^+ dominant, then τ_Λ is a minimal K-type of $\mathrm{ind}_{K\cap M^\#}^{K}(\sigma^\#)$.

Generically Δ_0 is equal to Δ_r, and then Theorem 2 below says that every Λ defined by (2.1) is automatically Δ_K^+ dominant; in this case the minimal τ_Λ's and the fine τ_μ's are in one-one correspondence. In the exceptional cases when $\Delta_0 \supsetneqq \Delta_r$, the fine μ's that lead to minimal Λ's are exactly those that satisfy certain conditions relative to the members of $\Delta_0 - \Delta_r$. The theorem uses the following notation: t_i denotes +1 or -1, α_i is a member of our strongly orthogonal set in $\Delta_{B,n}$, and ϵ is a member of $(t^{\mathbb{C}})'$ orthogonal to $(t_r^{\mathbb{C}})'$.

Theorem 2. If τ_μ is a fine K_r-type with $\tau_\mu|_{M_r}$ containing the character ω in (2.2), then the integral form Λ defined by (2.1) is Δ_K^+ dominant if and only if μ satisfies all of the following conditions:

(i) $2\langle\mu,\beta\rangle/|\beta|^2 > -1/2$ for each Δ_K^+ simple root β in $\Delta_0 - \Delta_r$ of the form $\beta = \epsilon - \frac{1}{2}t\alpha$ such that $|\beta| = |\alpha|$ and also $\epsilon - \frac{1}{2}\alpha$ and α are simple for Δ^+.

(ii) $2\langle\mu,\beta\rangle/|\beta|^2 > -1$ for each Δ_K^+ simple root β in $\Delta_0 - \Delta_r$ of the form $\beta = \varepsilon - \frac{1}{2}t_i\alpha_i - \frac{1}{2}t_j\alpha_j$ such that $|\beta| = |\alpha_i| = |\alpha_j|$, $\frac{1}{2}(\alpha_i + \alpha_j)$ is not in Δ, index i precedes index j for the ordering, $\varepsilon - \frac{1}{2}\alpha_i - \frac{1}{2}\alpha_j$ and α_j are simple for Δ^+, and either $t_i = 1$ or α_i is simple for Δ^+.

(iii) $2\langle\mu,\beta\rangle/|\beta|^2 > -1$ for each Δ_K^+ simple root β in $\Delta_0 - \Delta_r$ of the form $\beta = \varepsilon - \frac{1}{2}t_j\alpha_j$ such that $2|\beta|^2 = |\alpha_j|^2$, $\varepsilon - \frac{1}{2}\alpha_1$ is simple for Δ^+ when α_1 is the first α such that $\frac{1}{2}(\alpha + \alpha_j)$ is in Δ, and either $t_j = 1$ or α_j is simple for Δ^+.

Prototypes for the situations described in (i), (ii), and (iii) occur with the minimal parabolic subgroup of $SU(2,1)$ in the case of (i), the minimal parabolic subgroup of $SU(2,2)$ in the case of (ii), and the maximal parabolic subgroup of $Sp(2,\mathbb{R})$ with nonabelian N in the case of (iii). Case (iii) may be dropped from the theorem if $\sigma^{\#}$ is a genuine discrete series representation.

The proofs of the two theorems are straightforward but rather long. One proves the integrality first, and then the long step is Theorem 2. Next one constructs some μ satisfying the conditions in Theorem 2, and the rest is comparatively easy. We isolate from the proof one key lemma, which we shall use elsewhere.

<u>Lemma 3</u>. $2(\rho_K - \rho_{-,c}) = \rho - \rho_- - \rho_r + E(2\rho_K)$.

We conclude with some information about μ. It is always true that μ is a linear combination of the α_j's with coefficients 0, $\frac{1}{2}$, or $-\frac{1}{2}$. When G_r is locally a product of copies of $SL(2,\mathbb{R})$, i.e., when $\mathfrak{a} = 0$ and Δ_r is a product of root systems A_1, we can be more precise. This condition on G_r is satisfied, for example, whenever the restricted roots of G form a system of type $(BC)_n$.

For each α_j let ρ_{α_j} be half the sum of the roots in Δ whose inner product with α_j is > 0 and whose inner product with all other α_k is $= 0$.

<u>Theorem 4</u>. Suppose G_r is locally a product of copies of $SL(2,\mathbb{R})$. If τ_μ is a fine K_r-type with $\tau_\mu|_{M_r}$ containing the character ω in (2.2), then μ is of the form

$$\mu = \sum s_j\alpha_j, \quad s_j = \pm\tfrac{1}{2}, \tag{2.3}$$

with the sum extended over exactly those j for which

$$\chi(\gamma_{\alpha_j}) = (-1)^{2\langle\rho_{\alpha_j},\alpha_j\rangle/|\alpha_j|^2}. \tag{2.4}$$

Moreover, every choice of signs in (2.3) leads to another such μ.

There is a mnemonic for this result. To each α_j, §7 of [6] associates[5] a "Plancherel factor" μ_{σ,α_j}. When (2.4) holds, μ_{σ,α_j} is the product of a polynomial and a cotangent; when (2.4) fails, μ_{σ,α_j} is the product of a polynomial and a tangent. Consequently Theorem 4 says that each cotangent-type α_j contributes to the fine K_r-type μ in a pair of ways, via coefficients $s_j = \pm\frac{1}{2}$, while the tangent-type α_j's contribute uniquely via coefficient $s_j = 0$.

It is known from Theorem 12.6 of [6] that reducibility of $U(P,\sigma,0)$ arises when these Plancherel factors fail to vanish at the origin. Theorems 2 and 4 say that this same phenomenon accounts for multiple minimal K-types of $U(P,\sigma,0)$. When σ is a discrete series representation, Theorem 1.4 of Vogan [9] explains this correspondence.

---

[5] See also §10 and Corollary 12.5 of [6].

References

[1] I. N. Bernstein, I. M. Gelfand, and S. I. Gelfand, Models of representations of compact Lie groups, Func. Anal. and Its Appl. 9 (1975), 322-324.

[2] Harish-Chandra, Representations of semisimple Lie groups V, Proc. Nat. Acad. Sci. USA 40 (1954), 1076-1077.

[3] H. Hecht and W. Schmid, A proof of Blattner's conjecture, Inventiones Math. 31 (1975), 129-154.

[4] A. W. Knapp and B. Speh, The role of basic cases in classification: theorems about unitary representations applicable to SU(N,2), this volume.

[5] A. W. Knapp and G. Zuckerman, Classification theorems for representations of semisimple Lie groups, "Non-Commutative Harmonic Analysis," Springer-Verlag Lecture Notes in Math. 587 (1977), 138-159.

[6] A. W. Knapp and G. J. Zuckerman, Classification of irreducible tempered representations of semisimple groups, Ann. of Math. 116 (1982).

[7] R. P. Langlands, On the classification of irreducible representations of real algebraic groups, mimeographed notes, Institute for Advanced Study, 1973.

[8] W. Schmid, On the characters of the discrete series, Inventiones Math. 30 (1975), 47-144.

[9] D. A. Vogan, The algebraic structure of the representation of semisimple Lie groups I, Ann. of Math. 109 (1979), 1-60.

[10] D. A. Vogan, Fine K-types and the principal series, mimeographed notes, Massachusetts Institute of Technology, 1977.

Department of Mathematics
Cornell University
Ithaca, New York 14853, U.S.A.

THE ROLE OF BASIC CASES IN CLASSIFICATION:
Theorems about Unitary Representations Applicable to SU(N,2)

A. W. Knapp* and B. Speh*

We propose in this paper a nontrivial subdivision of the problem of classifying the irreducible unitary representations of semisimple Lie groups. It is known that the classification problem comes down to deciding which of certain standard representations induced from cuspidal parabolic subgroups and having a unique irreducible quotient admit a semidefinite inner product that makes the irreducible quotient unitary. The idea of the subdivision is to separate matters into a consideration of a small number of "basic cases" and a conjectural reduction step.

We confine ourselves to the situation that the underlying group G is linear and has rank equal to the rank of a maximal compact subgroup K. The standard representations that one has to consider are of the form

$$U(MAN,\sigma,\nu) = \operatorname{ind}_{MAN}^{G}(\sigma \otimes e^{\nu} \otimes 1)\,, \qquad (0.1)$$

where MAN is a cuspidal parabolic subgroup, σ is a discrete series or nondegenerate limit of discrete series representation of M, and ν is a real-valued linear functional on the Lie algebra of A in the closed positive Weyl chamber. (When ν is on the boundary of the Weyl chamber, some additional conditions are imposed on ν so that (0.1) has a unique irreducible quotient. See [11].)

---

* Supported by NSF Grant MCS 80-01854. The first author was supported also by Université Paris VII and by a Guggenheim Fellowship.

For each such MAN we shall give an existence result for certain σ's that we call "basic cases." We construct the basic cases explicitly when MAN is minimal. To any other σ we shall associate a proper reductive subgroup L with $\operatorname{rank} L = \operatorname{rank}(L \cap K)$ and a basic case σ^L of L, and a conjectural reduction step will describe the unitarity of the σ series in terms of the unitarity of the σ^L series. Part of the conjecture is closely related to conjectures by D. A. Vogan ([18], p. 408); the conjecture says also that L is large enough for a comparison of unitarity at all A parameters.

The discussion of the basic cases and the reduction conjecture are in §§1 and 3-5. They form the core of the paper. In §§2 and 6-10 we give a number of theorems that can be regarded as evidence for the conjecture or as treatment of basic cases. What these theorems have in common is that they all give new nontrivial information about unitarity in $SU(N,2)$. Some of these results have been announced by us earlier ([11], [8]).

Of particular interest are two general results in §§6-7. One way of viewing basic cases is as minimal elements under translation of the M parameter toward the walls of the Weyl chambers of G, in the sense of the appendix of [13]. It follows from Conjecture 5.1 that this operation must preserve unitarity, and we state such a result for MAN minimal as Theorem 6.1.

In §7 we address a consistency question for the conjecture in the situation that $\dim A = \text{real-rank}(L)$ and σ^L is trivial on the M of the derived group L' of L. In this situation the trivial representation of L' occurs for a certain parameter $\nu = \nu_0$, and parameters ν with $|\nu| > |\nu_0|$ cannot lead to unitary representations of L. According to Conjecture 5.1, parameters ν with $|\nu| > |\nu_0|$ should not give unitary representations of G either, and this we verify as Theorem 7.1.

Our notion of basic cases evolved from the theorems in §§6-7 mentioned above and from a detailed study of the groups SU(N,2) and Sp(n,1). We obtained the general definition and theorem for MAN minimal only afterward. Upon seeing our constructive proof when MAN is minimal, Vogan was able to give an existential proof that applies also when MAN is nonminimal. The conjectural reduction was adjusted to take into account some examples supplied by Vogan for MAN nonminimal. We are grateful to Vogan for his suggestions and for permission to include his existence proof. We are grateful also to Welleda Baldoni Silva for highlighting her results [1] about Sp(n,1) in various ways for us at our request so that we could guess what the basic cases are in Sp(n,1).

Contents

1. Basic cases, minimal MAN

To keep the ideas clear, we shall begin with the situation of a minimal parabolic subgroup MAN of G. We shall define "basic cases" as certain representations of the compact group M. To do so, we first introduce the notion of a <u>format</u>. Recall that we are assuming G is linear and $\operatorname{rank} G = \operatorname{rank} K$.

We shall use the notation of [9], which we summarize briefly and incompletely here. Let $\mathfrak{g}$ and $\mathfrak{k}$ be the Lie algebras of G and K, let $\mathfrak{b} \subseteq \mathfrak{k}$ be a compact Cartan subalgebra of $\mathfrak{g}$, and let Δ and Δ_K be the sets of roots of $(\mathfrak{g}^{\mathbb{C}}, \mathfrak{b}^{\mathbb{C}})$ and $(\mathfrak{k}^{\mathbb{C}}, \mathfrak{b}^{\mathbb{C}})$, respectively.

Fix a sequence $\alpha_1, \ldots, \alpha_\ell$ of strongly orthogonal noncompact members of Δ. In order to arrive at a minimal parabolic subgroup of G, we assume that

$$\ell = \text{real-rank}(G) . \tag{1.1}$$

Let

$$\mathfrak{b}_r = \sum \mathbb{R} i H_{\alpha_j}$$

$$\mathfrak{b}_- = \text{orthocomplement in } \mathfrak{b} \text{ to } \mathfrak{b}_r .$$

Anticipating a Cayley transform, we say that a root in Δ is <u>real</u> if it is carried on $\mathfrak{b}_r$, <u>imaginary</u> if it is carried on $\mathfrak{b}_-$, and <u>complex</u> otherwise. Let Δ_r and Δ_- be the subsets of real and imaginary roots, respectively. We construct a split subgroup G_r of G from $\mathfrak{b}$ and the members of Δ_r, and we let $K_r = K \cap G_r$ be its maximal compact subgroup. Let E be the orthogonal projection of $(i\mathfrak{b})'$ on $(i\mathfrak{b}_r)'$.

We build a Cayley transform $\underset{\sim}{c}$ from the roots $\alpha_1, \ldots, \alpha_\ell$ and use it as in [9] to form MA. If $\mathfrak{m}$ denotes the Lie algebra of M, we can regard Δ_- as the system of roots of $(\mathfrak{m}^{\mathbb{C}}, \mathfrak{b}_-^{\mathbb{C}})$. In [9] we defined

$$M^{\#} = M_e M_r , \tag{1.2}$$

where M_e is the analytic subgroup corresponding to $\mathfrak{m}$ and M_r is a finite abelian group built from the real roots. Because of (1.1), we have

$$M = M^{\#} . \tag{1.3}$$

Fix a positive system $(\Delta_-)^+$ for Δ_-. An irreducible representation σ of M is then determined by the highest weight λ of $\sigma|_{M_e}$ and a compatible character χ of M_r. By means of the ordered basis (1.2) in [9], we can introduce a positive system Δ^+ in which $\lambda + \rho_-$ is Δ^+ dominant, $(i\mathfrak{b}_r)'$ is spanned by the real simple roots, and some other conditions are satisfied. Let μ (in $(i\mathfrak{b}_r)'$) be the highest weight of a fine K_r-type whose restriction to M_r contains the translate of χ given by (2.2) of [9]. We shall say that λ has <u>format</u> $(\{\alpha_j\}, \Delta^+, \chi, \mu)$ if the linear form

$$\Lambda = \lambda - E(2\rho_K) + 2\rho_{K_r} + \mu \tag{1.4}$$

given in Theorem 1 of [9] is Δ_K^+ dominant; Theorem 2 of [9] provides checkable necessary and sufficient conditions on μ for deciding this dominance.

<u>Theorem 1.1</u>. Suppose the group G with rank G = rank K has $G^{\mathbb{C}}$ simply connected. Fix a format $(\{\alpha_j\}, \Delta^+, \chi, \mu)$ corresponding to a minimal parabolic MAN. Among all highest weights λ with this format, there is a unique one λ_b such that any other λ with this format has $\lambda - \lambda_b$ dominant for Δ^+ and G-integral.

We call λ_b (or the associated representation of M) the <u>basic case</u> for the format $(\{\alpha_j\}, \Delta^+, \chi, \mu)$. If $G^{\mathbb{C}}$ is not simply connected, the basic case can still be defined as a member of $(i\mathfrak{b}_-)'$ by taking it to be the basic case for that format in the covering group that has a simply connected complexification.

Our constructive proof of Theorem 1.1 is too long to give here. It consists in writing down a formula for λ_b and verifying all the properties with the aid of some of the lemmas used to prove the theorems of [9]. However, Vogan has given a short existence proof that does not attempt to derive the formula, and we can include that. We shall therefore give some examples, followed by the formula for λ_b, followed by Vogan's proof.

Examples.

1) G of real rank one. Denote the real positive root by α. The fine K_r-type μ is 0 or $+\frac{1}{2}\alpha$ or $-\frac{1}{2}\alpha$, and it determines χ. Often two different formats (one with $\mu = +\frac{1}{2}\alpha$ and one with $\mu = -\frac{1}{2}\alpha$) will lead to the same basic case.

a) $\widetilde{SO}(2n,1)$, $n \geq 2$. Here $M = \widetilde{SO}(2n-1)$. The basic cases (as representations of M) are the trivial representation and the spin representation.

b) $SU(n,1)$, $n \geq 2$. Here

$$M = \left\{ \begin{pmatrix} \omega & & \\ & e^{i\theta} & \\ & & e^{i\theta} \end{pmatrix},\ \omega \in U(n-1) \text{ and total determinant} = 1 \right\}.$$

The basic cases are $\sigma(\text{this}) = e^{ik\theta}$ with $|k| \leq n$.

c) $Sp(n,1)$, $n \geq 2$. Here $M = SU(2) \times Sp(n-1)$. The basic cases are $\sigma = (k \times \text{fundamental}) \otimes 1$ with $0 \leq k \leq 2n-1$.

d) Real form of F_4. Here $M = \widetilde{SO}(7)$. There are five basic cases—the trivial representation, the first three multiples of the spin representation, and the Cartan composition of the spin representation with the standard representation. In classical notation their highest weights are $(0,0,0)$, $(\frac{1}{2},\frac{1}{2},\frac{1}{2})$, $(1,1,1)$, $(\frac{3}{2},\frac{3}{2},\frac{3}{2})$, and $(\frac{3}{2},\frac{1}{2},\frac{1}{2})$.

2) $SU(N,2)$, $N \geq 3$. Here

$$M = \left\{ \begin{pmatrix} \omega & & & & \\ & e^{i\theta} & & & \\ & & e^{i\varphi} & & \\ & & & e^{i\varphi} & \\ & & & & e^{i\theta} \end{pmatrix}, \quad \omega \in U(N-2) \text{ and total determinant} = 1 \right\}.$$

The basic cases are $\sigma(\text{this}) = e^{i(m\theta + n\varphi)}$ with $|m| \leq N-1$ and $|n| \leq N-1$.

<u>Formula</u> for λ_b. We define λ_b by giving its inner product with each Δ^+ simple root β. Namely

$$\frac{2\langle \lambda_b, \beta \rangle}{|\beta|^2} = \begin{cases} 0 & \text{if } \beta \text{ real or imaginary} \\ -\dfrac{2\langle \rho_-, \beta\rangle}{|\beta|^2} + \text{correction}(\beta) & \text{if } \beta \text{ complex.} \end{cases} \quad (1.5)$$

Here correction(β) is always 0, $\frac{1}{2}$, or 1, depending on the form of β. Let ϵ denote a member of $(i\mathfrak{b}_-)'$. Then

$$\text{correction}(\beta) = \begin{cases} \dfrac{1}{2} & \text{if } \beta = \epsilon - \tfrac{1}{2}\alpha_j,\ |\beta| = |\alpha_j|, \text{ and } \mu \perp \alpha_j \\ \dfrac{1}{2}(1 - \operatorname{sgn}\langle \mu, \gamma\rangle) & \text{if } \beta = \epsilon - \tfrac{1}{2}\alpha_j,\ |\beta| = |\alpha_j|,\ \mu \not\perp \alpha_j, \text{ and a sign} \pm \text{ is fixed so that } \gamma = \epsilon \pm \tfrac{1}{2}\alpha_j \text{ is compact} \\ \dfrac{1}{2} + \dfrac{2\langle \mu, \beta\rangle}{|\beta|^2} & \text{if } \beta = \epsilon - \alpha_j \text{ with } |\beta|^2 = 2|\alpha_j|^2 \\ \dfrac{2\langle \mu, \gamma\rangle}{|\gamma|^2} & \text{if } \beta = \epsilon - \tfrac{1}{2}\alpha_i - \tfrac{1}{2}\alpha_j,\ |\beta| = |\alpha_i| = |\alpha_j|, \text{ and a sign} \pm \text{ is fixed so that } \gamma = \epsilon - \tfrac{1}{2}\alpha_i \pm \tfrac{1}{2}\alpha_j \text{ is compact.} \end{cases}$$

Proof of Theorem 1.1. Uniqueness is trivial. For existence, fix some λ corresponding to the format $(\{\alpha_j\},\Delta^+,\chi,\mu)$. By adding suitable fundamental weights for G to λ, we may assume $\lambda+\rho_-$ is nonsingular with respect to all nonreal roots. Let σ be the representation of M determined by λ and χ, and form the unitary principal series representation $U = U(MAN,\sigma,0)$. The μ in the format picks out one minimal K-type Λ of U by (1.4), and Theorem 1.1 of [17] says that Λ lies in a unique irreducible constituent π of U. For each non-real simple root β for Δ^+, let Λ_β be the fundamental weight corresponding to β and define n_β to be the greatest integer

$$n_\beta = \left[\frac{2\langle \lambda+\rho_-, \beta\rangle}{|\beta|^2}\right].$$

Then $\lambda+\rho_- - n_\beta\Lambda_\beta$ is Δ^+ dominant and we can apply the Zuckerman ψ functor [19] to π, obtaining

$$\psi^{\lambda+\rho_-}_{\lambda+\rho_- - n_\beta\Lambda_\beta}(\pi). \tag{1.6}$$

Define n'_β by

$$n'_\beta = \begin{cases} n_\beta & \text{if (1.6) is not } 0 \\ n_\beta - 1 & \text{if (1.6) is } 0. \end{cases}$$

Then

$$\psi^{\lambda+\rho_-}_{\lambda+\rho_- - n'_\beta\Lambda_\beta}(\pi)$$

is not 0. By Theorem 6.18 of Speh-Vogan [15],

$$\psi^{\lambda+\rho_-}_{\lambda+\rho_- - \Sigma n'_\beta\Lambda_\beta}(\pi) \tag{1.7}$$

is not 0. Put $\lambda_b = \lambda - \Sigma\, n'_\beta\Lambda_\beta$. Theorem B.1 of [13] shows that (1.7) is contained in an induced representation $U_b = U(MAN,\sigma_b,0)$

and that the infinitesimal character of σ_b is obtained by moving the infinitesimal character $\lambda+\rho_-$ of σ by $-\sum n'_\beta \Lambda_\beta$. The parameter χ is not changed. The highest weight of σ_b therefore has to be λ_b, and so λ_b is $(\Delta_-)^+$ dominant. We have arranged that $\lambda_b+\rho_-$ is Δ^+ dominant and that

$$0 \leq \frac{2\langle \lambda_b+\rho_-, \beta\rangle}{|\beta|^2} \leq 1.$$

The minimal K-type of (1.7) is easily seen to be $\Lambda - \sum n'_\beta \Lambda_\beta$.

Now let λ' correspond to the format $(\{\alpha_j\}, \Delta^+, \chi, \mu)$. It is not hard to see that $\lambda' - \lambda_b$ is G-integral. The only way that $\lambda' - \lambda_b$ can fail to be Δ^+ dominant is if some simple β_0 has

$$\frac{2\langle \lambda_b+\rho_-, \beta_0\rangle}{|\beta_0|^2} = 1 \quad \text{and} \quad \frac{2\langle \lambda'+\rho_-, \beta_0\rangle}{|\beta_0|^2} = 0. \tag{1.8}$$

So assume (1.8). We do not affect the definition of λ_b if we add enough fundamental weights for G to our initial λ so that $\lambda - \lambda'$ is Δ^+ dominant. The first equality in (1.8) implies that $n'_{\beta_0} = n_{\beta_0} - 1$. Hence (1.6) is 0 for $\beta = \beta_0$. The second equality in (1.8) allows us to compose (1.6) with a further ψ functor to obtain

$$\psi_{\lambda'+\rho_-}^{\lambda+\rho_-}(\pi) = 0. \tag{1.9}$$

If the form Λ' obtained by using λ' in (1.4) were Δ_K^+ dominant, we could construct a nonzero element in the space for (1.9) from the Λ K-type of π. Hence Λ' is not Δ_K^+ dominant, and (1.8) has led us to a contradiction.

2. Unitarity for some basic cases

We define the Langlands quotient

$$J(MAN,\sigma,\nu) \tag{2.1}$$

to be the unique irreducible quotient of (0.1) under the conditions on the (real-valued) element ν in the introduction. For the basic cases σ in the examples of §1, we shall describe those ν for which $J(MAN,\sigma,\nu)$ is infinitesimally unitary. Our description will be complete except for certain undecided isolated points in the case of $SU(N,2)$.

For groups of real rank one, the classification of irreducible unitary representations is known, with the final work appearing in [1] and [2]. In the notation of the examples of §1, denote the parameter on the Lie algebra of A by $\nu = t\,\underset{\sim}{c}(\alpha)$, $t > 0$, with the understanding that $\underset{\sim}{c}(\alpha)$ is a positive restricted root, and let ρ_A be half the sum of the positive restricted roots with multiplicities counted.

a) $\widetilde{SO}(2n,1)$, $n \geq 2$. Here ρ_A corresponds to $t = n - \frac{1}{2}$. For σ trivial, the unitary points are $0 < t \leq n - \frac{1}{2}$. For σ equal to the spin representation, no $t > 0$ gives a unitary point.

b) $SU(n,1)$, $n \geq 2$. Here ρ_A corresponds to $t = \frac{1}{2}n$. For $\sigma \leftrightarrow e^{ik\theta}$ with $|k| \leq n$, the unitary points are $0 < t \leq \frac{1}{2}(n - |k|)$.

c) $Sp(n,1)$, $n \geq 2$. Here ρ_A corresponds to $t = n + \frac{1}{2}$. For σ trivial, the unitary points are $0 < t \leq n - \frac{1}{2}$ and $t = n + \frac{1}{2}$. For $\sigma = (k \times \text{fundamental}) \otimes 1$ with $1 \leq k \leq 2n-1$, the unitary points are $0 < t \leq \frac{1}{2}(2n-1-k)$.

d) Real form of F_4. Here ρ_A corresponds to $t = \frac{11}{2}$. For σ trivial, the unitary points are $0 < t \leq \frac{5}{2}$ and $t = \frac{11}{2}$. For σ equal to k times the spin representation with $1 \leq k \leq 3$, the unitary points are $0 < t \leq \frac{1}{2}(3-k)$. For σ equal to the Cartan composition of the spin representation with the standard representation, no $t > 0$ gives a unitary point.

We turn to $SU(N,2)$, $N \geq 3$. With the notation for M as in §1, we have

$$A = \exp\begin{pmatrix} 0 & \begin{matrix} & & \end{matrix} \\ \begin{matrix} \\ \\ \\ \end{matrix} & \begin{matrix} & & s \\ & t & \\ t & & \\ s & & \end{matrix} \end{pmatrix},$$

and we let f_1 and f_2 of the Lie algebra matrix here be s and t, respectively. We write $\nu = af_1 + bf_2$ and choose the positive Weyl chamber to be $a \geq b \geq 0$. Then ρ_A has $a = N+1$ and $b = N-1$. The basic cases have $\sigma \leftrightarrow e^{i(m\theta+n\varphi)}$ with $|m| \leq N-1$ and $|n| \leq N-1$. Since complex conjugation is an outer automorphism of $SU(N,2)$ fixing A and sending (m,n) to $(-m,-n)$, it is enough to understand $m \geq n$.

Theorem 2.1. In $SU(N,2)$ the unitary points $\nu = (a,b)$ in the positive Weyl chamber for the basic cases $\sigma \leftrightarrow e^{i(m\theta+n\varphi)}$ with $m \geq n$ are as follows:

(a) If $|m| \leq N-2$ and $|n| \leq N-2$, the unitary points within the closed rectangle

$$0 \leq a \leq N-1-|m|, \quad 0 \leq b \leq N-1-|n| \tag{2.2}$$

are exactly the points

(i) in the triangle $a+b \leq m-n+2$

(ii) in any of the triangles

$$a-b \geq m-n+2k, \quad a+b \leq m-n+2k+2$$

for an integer $k \geq 1$.

(iii) on any of the lines

$$a - b = m - n + 2k$$

for an integer $k \geq 1$.

(b) If $|m| \leq N-2$ and $|n| \leq N-2$, the only possible unitary point outside the closed rectangle is

$$(a,b) = (N+1-|m|\,,\, N-1-|n|)\,. \qquad (2.3)$$

This can be a unitary point only if $m = n$ or $0 > m > n$. If $m = n = 0$, this point corresponds to the trivial representation.

(c) If $|m| = N-1$ or $|n| = N-1$ and if $m > n$, there are no unitary points in the positive Weyl chamber.

(d) If $|m| = N-1$ and $m = n$, the unitary points in the positive Weyl chamber are exactly the points $a + b \leq 2$.

This theorem will be proved in §10. Pictures of the unitary points for the cases $(m,n) = (0,0)$, $(0,-1)$, and $(2,2)$ appear in [11] and [8]. In situation (d) in the theorem, as well as in situation (c) when $n = -(N-1)$, the axis $b = 0$ is disallowed since $U(MAN,\sigma,\nu)$ does not have a unique irreducible quotient.

Vogan has shown us some computations indicating that at the undecided isolated points in (b) the representation $J(MAN,\sigma,\nu)$ has a highest weight vector. The unitary representations with a highest weight vector are known ([4] and [6]), and N. Wallach has given us information that suggests that these isolated points actually do correspond to unitary representations.

3. Basic cases, general MAN

Only minor modifications are needed to define "basic cases" for a general cuspidal parabolic subgroup MAN. The notation of [9] needs little adjustment. The system of strongly orthogonal noncompact roots $\{\alpha_j\}$ no longer need satisfy (1.1). The group $M^\#$ in (1.2) no longer need be all of M, but every discrete series or limit of discrete series σ of M is induced from a representation $\sigma^\#$ of the same type for $M^\#$.

Such a representation σ of M is therefore determined by a triple (λ_0, C, χ), where (λ_0, C) is a Harish-Chandra parameter of $\sigma^\#$ and χ is the scalar value of σ on $M_r \subseteq Z_M$. The chamber C for Δ_- determines $(\Delta_-)^+$, and the Blattner parameter (minimal $(K \cap M)$-type) of $\sigma^\#|_{M_e}$ is given by

$$\lambda = \lambda_0 - \rho_{-,c} + \rho_{-,n}. \tag{3.1}$$

We can define Δ^+ and μ as in §1, and we say λ_0 has <u>format</u> $(\{\alpha_j\}, \Delta^+, \chi, \mu)$ if the linear form Λ in (1.4) is Δ_K^+ dominant. (Observe that we have switched from the highest weight to the infinitesimal character λ_0 as reference parameter.)

<u>Theorem 3.1</u>. Suppose the group G with rank G = rank K has $G^{\mathbb{C}}$ simply connected. Fix a format $(\{\alpha_j\}, \Delta^+, \chi, \mu)$ corresponding to a general cuspidal parabolic subgroup MAN. Among all infinitesimal characters λ_0 of discrete series or limits (degenerate or nondegenerate) with this format, there is a unique one $\lambda_{b,0}$ such that any other λ_0 with this format has $\lambda_0 - \lambda_{b,0}$ dominant for Δ^+ and G-integral.

We call $\lambda_{b,0}$ (or its associated $\sigma = \sigma_b$) the <u>basic case</u> for the format $(\{\alpha_j\}, \Delta^+, \chi, \mu)$. Again $\lambda_{b,0}$ still makes sense if $G^{\mathbb{C}}$ is not simply connected.

The proof of Theorem 3.1 is the same as the proof we gave of Vogan's for Theorem 1.1 except for minor modifications. The σ_b that results is often, but not always, a limit of discrete series representation if MAN is not minimal. In fact, σ_b may even be degenerate as a limit of discrete series, and the theorem will fail if we look for $\lambda_{b,0}$ only among nondegenerate cases.

4. Associated subgroup L

Fix an infinitesimal character λ_0 for M and a format $(\{\alpha_j\},\Delta^+,\chi,\mu)$ for it. Let $\lambda_{b,0}$ be the basic case for this format given by Theorem 3.1, and let $\mathfrak{q} = \mathfrak{l}^{\mathbb{C}} \oplus \mathfrak{u}$ be the θ-stable parabolic subalgebra of $\mathfrak{g}^{\mathbb{C}}$ defined by the Δ^+ dominant form $\lambda_0 - \lambda_{b,0}$:

$$\begin{aligned} &\mathfrak{q} \quad \text{is built from } \mathfrak{b}^{\mathbb{C}} \text{ and all } \beta \in \Delta \text{ with } \langle \lambda_0 - \lambda_{b,0}, \beta\rangle \geq 0, \\ &\mathfrak{l}^{\mathbb{C}} \quad \text{is built from } \mathfrak{b}^{\mathbb{C}} \text{ and all } \beta \in \Delta \text{ with } \langle \lambda_0 - \lambda_{b,0}, \beta\rangle = 0, \\ &\mathfrak{u} \quad \text{is built from all } \beta \in \Delta \text{ with } \langle \lambda_0 - \lambda_{b,0}, \beta\rangle > 0. \end{aligned} \tag{4.1}$$

Here $\mathfrak{l}^{\mathbb{C}}$ is the complexification of $\mathfrak{l} = \mathfrak{l}^{\mathbb{C}} \cap \mathfrak{g}$. The analytic subgroup of G corresponding to $\mathfrak{l}$ will be denoted L; it is the centralizer in G of a suitable torus.

From the definition, $\mathfrak{b}^{\mathbb{C}}$ is contained in $\mathfrak{l}^{\mathbb{C}}$, and the root system of $(\mathfrak{l}^{\mathbb{C}},\mathfrak{b}^{\mathbb{C}})$ is

$$\Delta^L = \{\beta \in \Delta \mid \langle \lambda_0 - \lambda_{b,0}, \beta\rangle = 0\}.$$

Moreover, each α_j in our strongly orthogonal sequence of noncompact roots is in Δ^L. Hence $\mathfrak{g}_r \subseteq \mathfrak{l}$ and $A \subseteq L$. We shall associate to the Langlands quotient $J(MAN,\sigma,\nu)$ for G, given in (2.1), a Langlands quotient

$$J^L((M \cap L)A(N \cap L),\sigma^L,\nu) \tag{4.2}$$

for L. For brevity we shall write $J^G(\sigma,\nu)$ and $J^L(\sigma^L,\nu)$ for such a corresponding pair of representations of G and L.

To specify (4.2) we need to define σ^L, and we do so by giving its infinitesimal character λ_0^L and a compatible format

$$(\{\alpha_j\}, \Delta^+ \cap \Delta^L, \chi^L, \mu) . \tag{4.3}$$

With superscripts "L" referring to objects in L and to the positive system $\Delta^+ \cap \Delta^L$ for L, we define

$$\begin{aligned} \rho(\mathfrak{u}) &= \rho - \rho^L \\ \rho(\mathfrak{u} \cap \mathfrak{k}^{\mathbb{C}}) &= \rho_K - \rho_K^L \\ \rho(\mathfrak{u} \cap \mathfrak{p}^{\mathbb{C}}) &= \rho(\mathfrak{u}) - \rho(\mathfrak{u} \cap \mathfrak{k}^{\mathbb{C}}) . \end{aligned}$$

The form $\rho(\mathfrak{u})$ is orthogonal to every root in Δ^L, and $\rho(\mathfrak{u} \cap \mathfrak{k}^{\mathbb{C}})$ and $\rho(\mathfrak{u} \cap \mathfrak{p}^{\mathbb{C}})$ are orthogonal to every root in Δ_K^L.

Because of this orthogonality, $E(2\rho(\mathfrak{u} \cap \mathfrak{k}^{\mathbb{C}}))$ is the differential of a one-dimensional representation of K_r, and we can define

$$\chi^L = \chi \cdot [\exp E(2\rho(\mathfrak{u} \cap \mathfrak{k}^{\mathbb{C}}))]|_{M_r} . \tag{4.4}$$

Let

$$\lambda_0^L = \lambda_0 - \rho(\mathfrak{u}) . \tag{4.5}$$

These definitions are motivated by the theory of [15] and [18] in a way that we shall describe in §5.

Proposition 4.1. The definitions (4.4) and (4.5) of χ^L and λ_0^L consistently define σ^L, and (4.3) is a compatible format. The corresponding form (1.4) for σ^L is given by

$$\Lambda^L = \Lambda - 2\rho(\mathfrak{u} \cap \mathfrak{p}^{\mathbb{C}}) . \tag{4.6}$$

Proof. It is clear that λ_0^L is dominant for $(\Delta_-^L)^+$. Writing the formula of Lemma 3 of [9] for G and for L and subtracting, we have

$$2\rho(\mathfrak{u}\cap\mathfrak{k}^{\mathbb{C}}) - 2\rho_{-,c} + 2\rho^L_{-,c} = \rho(\mathfrak{u}) - \rho_- + \rho^L_- + E(2\rho(\mathfrak{u}\cap\mathfrak{k}^{\mathbb{C}})) .$$

Therefore the Blattner parameter (3.1) for λ_0^L can be transformed as

$$\begin{aligned}\lambda^L &= \lambda_0^L - \rho^L_{-,c} + \rho^L_{-,n}\\ &= [\lambda_0 - \rho(\mathfrak{u})] + [2\rho(\mathfrak{u}\cap\mathfrak{k}^{\mathbb{C}}) - E(2\rho(\mathfrak{u}\cap\mathfrak{k}^{\mathbb{C}})) - \rho(\mathfrak{u}) + \rho_{-,n} - \rho_{-,c}]\\ &= \lambda - 2\rho(\mathfrak{u}) + [2\rho(\mathfrak{u}\cap\mathfrak{k}^{\mathbb{C}}) - E(2\rho(\mathfrak{u}\cap\mathfrak{k}^{\mathbb{C}}))] . \qquad (4.7)\end{aligned}$$

Each of the terms on the right is analytically integral on $\exp \mathfrak{b}_-$, and hence λ^L is analytically integral on $\exp \mathfrak{b}_-$. We shall prove that

$$\langle \lambda^L , \beta\rangle = \langle \lambda , \beta\rangle \quad \text{for } \beta \in (\Delta^L_{-,c})^+ , \qquad (4.8)$$

and then it follows that λ_0^L is the infinitesimal character of a discrete series or limit of discrete series of $(M\cap L)_e$.

Thus let β be in $\Delta^L_{-,c}$. Then $\langle 2\rho(\mathfrak{u}) , \beta\rangle = 0$ since β is in Δ^L, and $\langle E(2\rho(\mathfrak{u}\cap\mathfrak{k}^{\mathbb{C}})) , \beta\rangle = 0$ since β is in Δ_-. Thus we are to show that

$$\langle 2\rho(\mathfrak{u}\cap\mathfrak{k}^{\mathbb{C}}) , \beta\rangle = 0 . \qquad (4.9)$$

If β is compact for $\mathfrak{g}^{\mathbb{C}}$, (4.9) is clear. If β is noncompact for $\mathfrak{g}^{\mathbb{C}}$, then β is orthogonal but not strongly orthogonal to some member α of the sequence $\{\alpha_j\}$. Then $\beta+\alpha$ and $\beta-\alpha$ are in Δ^L_K, and the product $s_{\beta+\alpha}s_{\beta-\alpha}$ of two reflections fixes $2\rho(\mathfrak{u}\cap\mathfrak{k}^{\mathbb{C}})$. But then

$$\begin{aligned}2\rho(\mathfrak{u}\cap\mathfrak{k}^{\mathbb{C}}) &= s_{\beta+\alpha}s_{\beta-\alpha}(2\rho(\mathfrak{u}\cap\mathfrak{k}^{\mathbb{C}}))\\ &= s_\alpha s_\beta(2\rho(\mathfrak{u}\cap\mathfrak{k}^{\mathbb{C}}))\\ &= 2\rho(\mathfrak{u}\cap\mathfrak{k}^{\mathbb{C}}) - \frac{2\langle 2\rho(\mathfrak{u}\cap\mathfrak{k}^{\mathbb{C}}),\alpha\rangle}{|\alpha|^2}\alpha - \frac{2\langle 2\rho(\mathfrak{u}\cap\mathfrak{k}^{\mathbb{C}}),\beta\rangle}{|\beta|^2}\beta ,\end{aligned}$$

and (4.9) follows. This proves (4.8).

Next we show that $\exp \lambda^L$ and χ^L agree on $(\exp \mathfrak{b}_-)\cap(\exp \mathfrak{b}_r)$, so that we obtain a well defined representation of $(M\cap L)^\#$, then of

$M \cap L$. In view of (4.7) and (4.4), we are to show that the character

$$\xi_{-2\rho(\mathfrak{u})}\xi_{[2\rho(\mathfrak{u}\cap\mathfrak{k}^{\mathbb{C}})-E(2\rho(\mathfrak{u}\cap\mathfrak{k}^{\mathbb{C}}))]} \tag{4.10}$$

of $\exp \mathfrak{b}_-$ and the character

$$\xi_{E(2\rho(\mathfrak{u}\cap\mathfrak{k}^{\mathbb{C}}))} \tag{4.11}$$

of $\exp \mathfrak{b}_r$ agree on $(\exp \mathfrak{b}_-) \cap (\exp \mathfrak{b}_r)$. The first factor of (4.10) is well defined on all of $\exp \mathfrak{b}$ and is trivial on $\exp \mathfrak{b}_r$. The second factor of (4.10) and the character (4.11) are the respective restrictions of the character $\xi_{2\rho(\mathfrak{u}\cap\mathfrak{k}^{\mathbb{C}})}$ of $\exp \mathfrak{b}$. The required consistency is therefore proved.

To see that (4.3) is a compatible format, we check that the representation τ_μ of K_r with highest weight μ contains

$$\chi^L \cdot \exp(E(2\rho_K^L) - 2\rho_{K_r})|_{M_r}. \tag{4.12}$$

Using (4.4), we see that (4.12) equals

$$\chi \cdot \exp(E(2\rho_K) - 2\rho_{K_r})|_{M_r},$$

and τ_μ contains this by assumption.

Finally we combine (1.4) and Lemma 3 of [9] to write

$$\Lambda = \lambda_0 + \rho - \rho_r - 2\rho_K + 2\rho_{K_r} + \mu. \tag{4.13}$$

Writing the corresponding expression for Λ^L and subtracting, we have

$$\begin{aligned}\Lambda^L - \Lambda &= (\lambda_0^L - \lambda_0) - (\rho - \rho^L) + (2\rho_K - 2\rho_K^L)\\ &= -\rho(\mathfrak{u}) - \rho(\mathfrak{u}) + 2\rho(\mathfrak{u}\cap\mathfrak{k}^{\mathbb{C}})\\ &= -2\rho(\mathfrak{u}\cap\mathfrak{p}^{\mathbb{C}}).\end{aligned}$$

This proves (4.6) and completes the proof of the proposition.

The group L is reductive, not necessarily semisimple, and we have to adjust the definitions of §§1-3 to speak of "basic cases" for L. Let us agree that a basic case for L is one whose restriction to the semisimple part of L is basic. In terms of a comparison of infinitesimal characters with a given format, one therefore fixes the restriction to the central torus of L.

Proposition 4.2. The infinitesimal character λ_0^L given in (4.5) is a basic case for the format (4.3) for L.

Proof. We may assume that $G^{\mathbb{C}}$ is simply connected. Proposition 4.1 shows that λ_0^L does correspond to a nonzero representation with (4.3) as format. Suppose λ_0^L is not a basic case, i.e., that there is some Δ^+ dominant integral ξ not orthogonal to Δ^L such that $\lambda_0^L - \xi$ corresponds to a nonzero representation with (4.3) as format. For each Δ^+ simple β in Δ outside Δ^L, let Λ_β be the fundamental weight, and let η be the sum of such Λ_β. Then we claim that

$$\lambda_0' = \lambda_0 + n\eta - \xi$$

corresponds to a nonzero representation with $(\{\alpha_j\},\Delta^+,\chi,\mu)$ as format, provided n is sufficiently large.

In fact, the integrality condition is no problem. The other conditions are that certain inner products of λ_0' or its translates with certain members of Δ^+ are to be ≥ 0. When these members are in Δ^L, we have the same inner product as for $\lambda_0^L - \xi$. When they are outside Δ^L, the n dominates (if n is sufficiently large) and makes the inner product ≥ 0.

Now choose β simple for Δ^L so that $\langle\xi,\beta\rangle > 0$. Then

$$\langle\lambda_0' - \lambda_{b,0}, \beta\rangle = \langle\lambda_0 - \lambda_{b,0}, \beta\rangle - \langle\xi, \beta\rangle$$
$$= - \langle\xi, \beta\rangle < 0,$$

and we have a contradiction to the fact that $\lambda_{b,0}$ is basic for G. The proposition follows.

5. Conjecture about reduction

Conjecture 5.1. Let σ be a discrete series or nondegenerate limit of discrete series representation of M, given by the infinitesimal character λ_0 and a compatible format $(\{\alpha_j\},\Delta^+,\chi,\mu)$. Let L be defined from (4.1), and let σ^L be defined as in (4.4) and (4.5). If ν is real-valued, then the Langlands quotient $J^G(\sigma,\nu)$ for G is infinitesimally unitary if and only if the Langlands quotient $J^L(\sigma^L,\nu)$ for L is infinitesimally unitary.

If true, the conjecture reduces the classification question for irreducible unitary representations to a consideration of $J(\sigma,\nu)$ with σ basic and ν real-valued, under the assumption that G is linear and rank G = rank K. (This follows from Proposition 4.2.)

The conjecture is true if G has real rank one. For $G = SU(N,2)$, there are two proper cuspidal parabolic subgroups MAN to consider: When dim A = 1, the conjecture is true at least when σ is a discrete series representation of M and L has real rank one; this follows essentially from Proposition 9.1 below. When dim A = 2, the conjecture is true inside a certain rectangle of ν's (by Propositions 8.1 and 8.2 below), and it is often true also outside a slightly larger circle of ν's (cf. Theorem 7.1), but in general the conjecture is not settled.

Actually the conjecture should be regarded as suggesting more than just a correspondence of unitary parameters. It should suggest that certain functors going back and forth between representations of L and representations of G preserve unitarity. These functors in many situations coincide with the ones of Vogan ([17],[18]) and

Speh and Vogan [15], and the choice of parameters in (4.4), (4.5), and (4.6) is in fact motivated by the Vogan and Speh-Vogan functors.

The correspondence between our work and that in [15], [17], and [18] is not immediately evident, since the orderings are different. The ordering in [17] is obtained by associating to the minimal K-type Λ a form $\tilde{\lambda}$ via Proposition 4.1 of [17]. Then one builds a smallest permissible θ-stable parabolic subalgebra for the theory from $\{\beta \in \Delta \mid \langle \tilde{\lambda}, \beta \rangle \geq 0\}$. The point is that $\tilde{\lambda}$ is just the infinitesimal character λ_0, so that our $\mathfrak{q}$ is indeed a permissible θ-stable parabolic subalgebra.

It is implicit in §7 of [17] that $\tilde{\lambda} = \lambda_0$, and we shall here write out a direct proof. From (4.13) we have

$$\Lambda + 2\rho_K = \lambda_0 + (\rho - \rho_r) + (\mu + 2\rho_{K_r}) . \tag{5.1}$$

We introduce a new positive system $(\Delta^+)'$ by changing the notion of positivity on Δ_r (and only there) so that $\mu + 2\rho_{K_r}$ is $(\Delta_r^+)'$ dominant. With "primes" referring to objects in the new ordering, we have

$$\Delta^+_{K,r} = (\Delta^+_{K,r})' \quad \text{and} \quad \Delta^+_K = (\Delta^+_K)'$$

$$\rho_{K_r} = \rho'_{K_r} \quad \text{and} \quad \rho_K = \rho'_K$$

$$\rho - \rho_r = \rho' - \rho'_r . \tag{5.2}$$

By Proposition 4.1 of [17], we can write

$$\mu + 2\rho_{K_r} - \rho'_r = \tilde{\lambda}_\mu - \tfrac{1}{2}\Sigma\, \beta_i , \tag{5.3}$$

where the β_i are in $(\Delta_r^+)'$ and have certain properties listed in the proposition. Since μ is fine and "fine" is equivalent with "small" ([18], p. 294), $\tilde{\lambda}_\mu = 0$. Combining (5.1), (5.2), and (5.3), we obtain

$$\Lambda + 2\rho_K - \rho' = \lambda_0 - \tfrac{1}{2}\Sigma\,\beta_i . \qquad (5.4)$$

The claim is that $\Lambda + 2\rho_K$ is dominant for $(\Delta^+)'$ and hence that $\tilde{\lambda} = \lambda_0$ by the uniqueness in Proposition 4.1 of [17].

In fact, let β be simple for $(\Delta^+)'$. If β is not in $(\Delta_r^+)'$, then $\langle\beta,\rho'\rangle \geq 0$ and $\langle\beta,\beta_i\rangle \leq 0$ for all i. Also β is in Δ^+, and hence $\langle\lambda_0,\beta\rangle \geq 0$. Thus $\langle\Lambda + 2\rho_K, \beta\rangle \geq 0$ by (5.4). On the other hand, if β is in $(\Delta_r^+)'$, then β is orthogonal to the first two terms on the right of (5.1) and has inner product ≥ 0 with the third term by construction. Thus $\langle\Lambda + 2\rho_K, \beta\rangle \geq 0$.

Hence $\tilde{\lambda} = \lambda_0$, and our $\mathfrak{q}$ is permissible in the theory of [15], [17], and [18]. Conjecture 5.1 is thus closely related to the two conjectures on page 408 of [18]. One additional thing that Conjecture 5.1 says is that our L is large enough to capture all the unitary points in G.

6. Preservation of unitarity under tensoring

In [19] and the appendix of [13], G. Zuckerman began a systematic investigation of one technique for moving a parameter by a discrete step through a series of representations of a connected semisimple Lie group. The technique consists of tensoring with a suitable finite-dimensional representation and projecting according to a particular value of the infinitesimal character. This is done in two distinct ways—by a ψ functor that makes the parameter smaller and by a φ functor that makes the parameter larger.

Since finite-dimensional representations are generally not unitary, this technique need not carry unitary representations to unitary representations in general. However, Conjecture 5.1 predicts that unitarity will be preserved for Langlands quotients when ψ or

φ moves only the M parameter. (For example, $\lambda_0 - \lambda_{b,0}$ is the highest weight of a finite-dimensional representation that moves the parameter this way.) The point of this section will be to verify this prediction under the additional assumption that MAN is minimal.

Let G be a linear connected semisimple group with maximal compact subgroup K. For this section only, we do not assume rank G = rank K. Let MAN be a minimal parabolic subgroup of G, and let $\mathfrak{g}$, $\mathfrak{k}$, $\mathfrak{m}$, $\mathfrak{a}$, and $\mathfrak{n}$ be the various Lie algebras corresponding to our Lie groups. Let $\mathfrak{b}_- \subseteq \mathfrak{m}$ be a maximal abelian subspace (so that $\mathfrak{a} \oplus \mathfrak{b}_-$ is a Cartan subalgebra of $\mathfrak{g}$), let $B_- = \exp \mathfrak{b}_-$, let Δ be the roots of $(\mathfrak{g}^{\mathbb{C}}, (\mathfrak{a} \oplus \mathfrak{b}_-)^{\mathbb{C}})$, and let $\Delta_- \subseteq \Delta$ be the roots of $(\mathfrak{m}^{\mathbb{C}}, \mathfrak{b}_-^{\mathbb{C}})$. A positive system Δ^+ for Δ will be said to be compatible with a positive system $(\Delta_-)^+$ for Δ_- if $(\Delta_-)^+ \subseteq \Delta^+$. (No compatibility of Δ^+ with $\mathfrak{n}$ is assumed.)

If $(\Delta_-)^+$ is specified, let ρ_- denote half the sum of the members of $(\Delta_-)^+$. An irreducible unitary (finite-dimensional) representation σ of M is determined by a pair (λ, χ), where

λ is a dominant analytically integral member of $(i\mathfrak{b}_-)'$

χ is a character of $M \cap \exp i\mathfrak{a}$ that agrees with e^{λ} on $B_- \cap \exp i\mathfrak{a}$,

the correspondence being that λ is the highest weight of $\sigma|_{M_e}$ and χ is the scalar value of σ on $M \cap \exp i\mathfrak{a} \subseteq Z_M$. We call χ the central character of σ.

Suppose ν in $(\mathfrak{a}')^{\mathbb{C}}$ has $\operatorname{Re} \nu$ in the closed positive Weyl chamber of $\mathfrak{a}'$ determined by $\mathfrak{n}$. (We do not need to assume ν is real-valued.) Theorem 1.1 of [11] recalls a necessary and sufficient condition for the induced representation $U(MAN, \sigma, \nu)$ to have a unique irreducible quotient, which we define to be $J(MAN, \sigma, \nu)$. For an application in Proposition 8.2, we note now that $J(MAN,\sigma,\nu)$ is always defined if $\mathrm{Re}\,\nu$ is in the interior of the positive Weyl chamber.

The two theorems to follow concern the effect on unitarity of moving the σ parameter of $J(MAN,\sigma,\nu)$. In the notation of [19], Theorem 6.1 deals with the ψ functor and Theorem 6.2 deals with the φ functor.

Theorem 6.1. Let MAN be a minimal parabolic subgroup of G, fix a positive system $(\Delta_-)^+$ for M, and let σ and σ' be irreducible unitary representations of M with respective highest weights λ and λ' and with a common central character χ. Let ν be in $(\mathfrak{a}')^{\mathbb{C}}$ with $\mathrm{Re}\,\nu$ in the closed positive Weyl chamber, and let Δ^+ be a positive system of roots of $(\mathfrak{g}^{\mathbb{C}}, (\mathfrak{a}\oplus\mathfrak{b}_-)^{\mathbb{C}})$ compatible with $(\Delta_-)^+$. Suppose that

(i) $J(MAN,\sigma',\nu)$ is defined

(ii) $\lambda'+\rho_-+\mathrm{Re}\,\nu$ and $\lambda+\rho_-+\mathrm{Re}\,\nu$ are Δ^+ dominant

(iii) $\lambda'-\lambda$ is Δ^+ dominant and G-integral.

Then

(iv) $J(MAN,\sigma,\nu)$ is defined

(v) $J(MAN,\sigma',\nu)$ infinitesimally unitary implies $J(MAN,\sigma,\nu)$ infinitesimally unitary.

Theorem 6.2. Let MAN be a minimal parabolic subgroup of G, fix a positive system $(\Delta_-)^+$ for M, and let σ and σ' be irreducible unitary representations of M with respective highest weights λ and λ' and with a common central character χ. Let ν be in $(\mathfrak{a}')^{\mathbb{C}}$ with $\mathrm{Re}\,\nu$ in the closed positive Weyl chamber, and let Δ^+ be a positive system of roots of $(\mathfrak{g}^{\mathbb{C}}, (\mathfrak{a}\oplus\mathfrak{b}_-)^{\mathbb{C}})$ compatible with $(\Delta_-)^+$. Suppose that

(i) $J(MAN,\sigma',\nu)$ is defined

(ii) $\lambda' + \rho_- + \mathrm{Re}\,\nu$ and $\lambda + \rho_- + \mathrm{Re}\,\nu$ are Δ^+ dominant and are equisingular (i.e., singular with respect to the same roots of $(\mathfrak{g}^{\mathbb{C}}, (\mathfrak{a} \oplus \mathfrak{b}_-)^{\mathbb{C}})$)

(iii) $\lambda - \lambda'$ is Δ^+ dominant and G-integral.

Then

(iv) $J(MAN,\sigma,\nu)$ is defined

(v) $J(MAN,\sigma',\nu)$ infinitesimally unitary implies $J(MAN,\sigma,\nu)$ infinitesimally unitary.

These theorems will be proved on another occasion. Each proof consists in tracking down what happens to the relevant intertwining operator and seeing that positivity is preserved. From Theorem 6.1 and the results for the basic cases listed in Theorem 2.1, we can exclude many representations in $SU(N,2)$ from being unitary; we state a precise result in this direction as Proposition 8.2.

7. Zuckerman triples

In this section we give a general theorem applicable when G is linear and $\mathrm{rank}\, G = \mathrm{rank}\, K$ that says that $J(MAN,\sigma,\nu)$ cannot be infinitesimally unitary for real ν outside a certain radius, for a wide class of σ. Motivation for the theorem in terms of a construction of Zuckerman appears in [8] and will not be repeated here.

Thus let $\mathfrak{b} \subseteq \mathfrak{k}$ be a compact Cartan subalgebra of $\mathfrak{g}$, and let Δ be the set of roots of $(\mathfrak{g}^{\mathbb{C}}, \mathfrak{b}^{\mathbb{C}})$. We say that (Δ^+,Σ,χ) is a Zuckerman triple if

Δ^+ = a positive root system for Δ

Σ = a root system in Δ generated by Δ^+ simple roots

χ = an analytically integral form on $\mathfrak{b}^{\mathbb{C}}$ orthogonal to Σ with $\chi - \rho_\Delta + 2\rho_\Sigma$ dominant for Δ^+.

Let Δ_K and Σ_K be the subsystems of compact roots in Δ and Σ. We let $\rho_{\Delta,K}$, $\rho_{\Delta,n}$, $\rho_{\Sigma,K}$, and $\rho_{\Sigma,n}$ denote the half sums of the indicated positive compact or noncompact roots, and we let w_Σ and $w_{\Sigma,K}$ denote the long elements of the Weyl groups of Σ and Σ_K, respectively. We say that (Δ^+,Σ,χ) is <u>nondegenerate</u> if $\chi - \rho_\Delta + 2\rho_\Sigma$ is nonorthogonal to every root $w_\Sigma\beta$ with β in Δ_K.

<u>Theorem 7.1</u>. With rank G = rank K, suppose that a Langlands quotient $J(P,\sigma,\nu_0)$ is such that there is a Zuckerman triple (Δ^+,Σ,χ) for which $J(P,\sigma,\nu_0)$ has the real infinitesimal character $\chi - \rho_\Delta + 2\rho_\Sigma$ and a (Δ_K^+ dominant) minimal K-type $\chi - 2\rho_{\Delta,K} + 2\rho_{\Sigma,K}$. Suppose further that (Δ^+,Σ,χ) is nondegenerate; this condition is satisfied in particular if $\chi - \rho_\Delta + 2\rho_\Sigma$ is nonsingular. Then $J(P,\sigma,\nu)$ is not infinitesimally unitary for any real ν with $|\nu_0| < |\nu|$.

Remarks. This theorem was our first clue about basic cases. Its relevance is as follows: Under the assumptions in the theorem if also $\chi - \rho_\Delta + 2\rho_\Sigma$ is nonsingular, then the group L attached to σ by §4 often has Δ^L essentially equal to Σ.

The proof will use the Dirac inequality in the following form. See §4 of Baldoni Silva [1] for a proof of this inequality.

<u>Lemma 7.2</u>. If Δ^+ is any positive system for Δ and if π is an irreducible unitary representation of G with real infinitesimal character $\chi(\pi)$ and a minimal K-type Λ, then

$$|\lambda(\pi)| \leq |w(\Lambda - \rho_{\Delta,n}) + \rho_{\Delta,K}| , \tag{7.1}$$

where w is chosen in the Weyl group of Δ_K to make $w(\Lambda - \rho_{\Delta,n})$ be Δ_K^+ dominant.

Proof of Theorem 7.1. It is enough to prove that equality holds in (7.1) for $\pi = J(MAN,\sigma,\nu_0)$. We proceed in several steps.

(1) $w_{\Sigma,K}$ fixes $\chi - 2\rho_{\Delta,K} + 2\rho_{\Sigma,K}$.

In fact, we have

$$\begin{aligned} w_{\Sigma,K}\rho_{\Delta,K} &= \rho_{\Delta,K} - 2\rho_{\Sigma,K} \\ w_{\Sigma,K}\rho_{\Sigma,K} &= -\rho_{\Sigma,K} \\ w_{\Sigma,K}\chi &= \chi , \end{aligned} \tag{7.2}$$

and (1) follows.

(2) $w_{\Sigma,K}w_{\Sigma}(\chi - \rho_{\Delta} + 2\rho_{\Sigma}) = (\chi - 2\rho_{\Delta,K} + 2\rho_{\Sigma,K}) - w_{\Sigma,K}\rho_{\Delta,n} + \rho_{\Delta,K}$.

In fact, the left side is

$$\begin{aligned} &= \chi - w_{\Sigma,K}w_{\Sigma}(\rho_{\Delta} - \rho_{\Sigma}) + w_{\Sigma,K}w_{\Sigma}\rho_{\Sigma} \\ &= \chi - w_{\Sigma,K}(\rho_{\Delta} - \rho_{\Sigma}) - w_{\Sigma,K}\rho_{\Sigma} \\ &= \chi - w_{\Sigma,K}\rho_{\Delta} \\ &= \chi - w_{\Sigma,K}\rho_{\Delta,K} - w_{\Sigma,K}\rho_{\Delta,n} \\ &= \chi - \rho_{\Delta,K} + 2\rho_{\Sigma,K} - w_{\Sigma,K}\rho_{\Delta,n} \quad \text{by (7.2)} \\ &= (\chi - 2\rho_{\Delta,K} + 2\rho_{\Sigma,K}) - w_{\Sigma,K}\rho_{\Delta,n} + \rho_{\Delta,K}. \end{aligned}$$

(3) $|\chi - \rho_{\Delta} + 2\rho_{\Sigma}| = |w_{\Sigma,K}((\chi - 2\rho_{\Delta,K} + 2\rho_{\Sigma,K}) - \rho_{\Delta,n}) + \rho_{\Delta,K}|$.

In fact, we can take the magnitude of both sides of (2) and apply (1).

(4) $\langle w_{\Sigma,K}((\chi - 2\rho_{\Delta,K} + 2\rho_{\Sigma,K}) - \rho_{\Delta,n}), \beta\rangle \geq 0$ for $\beta \in \Sigma_K^+$.

In fact, the left side by (1) is

$$= \langle \chi - 2\rho_{\Delta,K} + 2\rho_{\Sigma,K}, \beta\rangle + \langle \rho_{\Delta,n}, -w_{\Sigma,K}\beta\rangle$$

$\geq \langle \rho_{\Delta,n}, -w_{\Sigma,K}\beta\rangle$ by the assumed Δ_K^+ dominance

≥ 0

since $-w_{\Sigma,K}\beta$ is in Σ_K^+ and $\rho_{\Delta,n}$ is Δ_K^+ dominant.

(5) $\langle w_{\Sigma,K}((\chi - 2\rho_{\Delta,K} + 2\rho_{\Sigma,K}) - \rho_{\Delta,n}), \beta\rangle \geq 0$ for $\beta \in \Delta_K^+$.

In fact, we may assume that β is Δ_K^+ simple, and by (4) we may assume β is not in Σ_K^+. By (1) and (2), we have

$$\langle w_{\Sigma,K}((\chi - 2\rho_{\Delta,K} + 2\rho_{\Sigma,K}) - \rho_{\Delta,n}) + \rho_{\Delta,K}, \beta\rangle$$
$$= \langle \chi - \rho_\Delta + 2\rho_\Sigma, w_\Sigma w_{\Sigma,K}\beta\rangle . \qquad (7.3)$$

Since β is in Δ_K^+ but not Σ, $w_{\Sigma,K}\beta$ is in Δ_K^+ but not Σ. Then $w_\Sigma(w_{\Sigma,K}\beta)$ is in Δ^+, and the dominance of $\chi - \rho_\Delta + 2\rho_\Sigma$ implies that (7.3) is ≥ 0. Since $w_{\Sigma,K}\beta$ is in Δ_K^+, the assumed nondegeneracy implies (7.3) is $\neq 0$. Therefore

$$\frac{2\langle w_{\Sigma,K}((\chi - 2\rho_{\Delta,K} + 2\rho_{\Sigma,K}) - \rho_{\Delta,n}) + \rho_{\Delta,K}, \beta\rangle}{|\beta|^2} \geq 1.$$

Since β is simple for Δ_K^+, $2\langle \rho_{\Delta,K}, \beta\rangle/|\beta|^2 = 1$. Then (5) follows.

(6) Comparing (3) and (5) with the statement of Lemma 7.2, we see that equality holds in (7.1) for $J(MAN,\sigma,\nu_0)$, and the theorem is proved.

Remark. For any nondegenerate Zuckerman triple, it is automatic that $\chi - 2\rho_{\Delta,K} + 2\rho_{\Sigma,K}$ is Δ_K^+ dominant. In fact, if β is in Σ_K^+, then $\langle \chi - 2\rho_{\Delta,K} + 2\rho_{\Sigma,K}, \beta\rangle = 0$. If β is in Δ_K^+ but not Σ_K, we have

$$\begin{aligned}\langle \chi - 2\rho_{\Delta,K} + 2\rho_{\Sigma,K}, \beta\rangle &= \langle \chi - 2\rho_{\Delta,K} + 2\rho_{\Sigma,K} - w_{\Sigma,K}\rho_{\Delta,n}, \beta\rangle + \langle w_{\Sigma,K}\rho_{\Delta,n}, \beta\rangle \\ &= \langle \chi - 2\rho_{\Delta,K} + 2\rho_{\Sigma,K} - w_{\Sigma,K}\rho_{\Delta,n}, \beta\rangle + \langle \rho_{\Delta,n}, \beta'\rangle \\ &\geq \langle \chi - 2\rho_{\Delta,K} + 2\rho_{\Sigma,K} - w_{\Sigma,K}\rho_{\Delta,n}, \beta\rangle ,\end{aligned}$$

and this is ≥ 0 by the same calculation as in (5).

8. Unitary degenerate series

We turn now to results that we shall formulate specifically only for $SU(N,2)$, $N \geq 3$. In this section we shall identify some unitary representations attached to the minimal parabolic subgroup. The new ones will be degenerate series, induced from a finite-dimensional representation of a noncuspidal maximal parabolic subgroup, and they have the striking feature that the finite-dimensional representation of the M of the maximal parabolic is usually nonunitary.

For $SU(N,2)$ with $N \geq 3$, we have already fixed a choice of the M and the A of a minimal parabolic subgroup in §§1-2, and we defined linear functionals f_1 and f_2 on the Lie algebra of A. We continue to write $\nu = af_1 + bf_2$, and we return to the assumption that ν is real-valued. The positive Weyl chamber is given by $a \geq b \geq 0$. An irreducible representation σ of M can be written (nonuniquely) as

$$\sigma\begin{pmatrix} \omega & & & & \\ & e^{i\theta} & & & \\ & & e^{i\varphi} & & \\ & & & e^{i\varphi} & \\ & & & & e^{i\theta} \end{pmatrix} = e^{i(m\theta + n\varphi)}\sigma_0(\omega) ,$$

where σ_0 is an irreducible representation of $U(N-2)$. If σ_0 has highest weight $\sum_{j=1}^{N-2} c_j e_j$, then the infinitesimal character of $U(MAN,\sigma,\nu)$ is

$$\sum_{j=1}^{N-2} (c_j + \tfrac{1}{2}(N-2j-1))e_j + \tfrac{m}{2}(e_{N-1} + e_{N+2}) + \tfrac{n}{2}(e_N + e_{N+1})$$
$$+ \tfrac{a}{2}(e_{N-1} - e_{N+2}) + \tfrac{b}{2}(e_N - e_{N+1}) . \quad (8.1)$$

We shall define a "fundamental rectangle" in the ν space. If we restrict σ to the subgroup of M where $\varphi = 0$, we obtain a representation σ_1 of the M for a subgroup $SU(N-1, 1)$ of $SU(N,2)$. The corresponding A for this $SU(N-1, 1)$ has $b = 0$. Let a_0 be the first point ≥ 0 such that the infinitesimal character of the representation of $SU(N-1, 1)$ induced from σ_1 and af_1 is integral and fails to be singular with respect to two linearly independent roots. Operationally a_0 is the first value ≥ 0 of a in $n-1-m+2\mathbb{Z}$ such that $\tfrac{1}{2}(m+a)$ and $\tfrac{1}{2}(m-a)$ do not both appear among the numbers $c_j + \tfrac{1}{2}(N-2j-1)$ for $1 \leq j \leq N-2$. Similarly the condition $\theta = 0$ leads us to a different subgroup $SU(N-1, 1)$ and to a representation σ_2 of its M, and we define b_0 relative to σ_2 and bf_2. The <u>fundamental rectangle</u> is then given by

$$0 \leq a \leq a_0 \quad \text{and} \quad 0 \leq b \leq b_0 .$$

In this section we shall identify some points in the fundamental rectangle that correspond to unitary representations. In Proposition 8.2 we shall see that the remaining points in the fundamental rectangle do not correspond to unitary representations. Because of Theorem 2.1, Conjecture 5.1 would imply that there is at most one ν (for fixed σ) outside the fundamental rectangle that corresponds to a unitary representation unless $a_0 = b_0 = 0$ and $m = n$.

If $a_0 = 0$ or $b_0 = 0$, then there are no unitary points at all unless $m = n$ (cf. [1], Theorem 6.1), and in this case the points with $b = 0$ do not have well defined Langlands quotients. Thus we shall assume $a_0 > 0$ and $b_0 > 0$ in our analysis.

We can determine which $U(MAN,\sigma,\nu)$ are reducible as in [10] by decomposing the standard intertwining operator for the large element of the 8-element Weyl group. The result is that the only reducibility within the fundamental rectangle occurs on the lines

$$a + b = |m - n| + 2\ell, \quad \ell \text{ an integer} \geq 1$$

$$a - b = |m - n| + 2k, \quad k \text{ an integer} \geq 1.$$

In view of Proposition 3.1 of [11], the representation $U(MAN,\sigma,\nu)$ at $b = 0$ with $0 \leq a \leq a_0$ is unitarily induced from a complementary series of $SU(N-1,1)$ and hence is unitary. Since the standard intertwining operator is the identity at $\nu = 0$, it follows from a familiar continuity argument that the following ν's in the positive Weyl chamber within the fundamental rectangle correspond to unitary representations:

(i) the triangle $\quad a + b \leq |m - n| + 2 \qquad (8.2)$

(ii) the triangles

$$a - b \geq |m - n| + 2k, \quad a + b \leq |m - n| + 2k + 2 \qquad (8.3)$$

for each integer $k \geq 1$.

These unitary points had been recognized earlier. (Cf. Knapp-Stein [12] for (i) and Guillemonat [5] for (ii).) Further unitary points are given in the following proposition, which was announced in [11]. Some of these points were recognized independently by Schlichtkrull [14].

Proposition 6.1. Within the fundamental rectangle when $a_0 > 0$ and $b_0 > 0$, the points ν on the lines

$$a - b = |m - n| + 2k , \quad k \text{ an integer} \geq 1, \tag{8.4}$$

correspond to unitary representations.

Proof. On any line (8.4), the argument with the intertwining operator that detected reducibility of $U(MAN,\sigma,\nu)$ shows also, just as in [10], that the Langlands quotient is an irreducible degenerate series representation as long as ν is in the interior of the fundamental rectangle and ν is not at a point where the line (8.4) crosses a line

$$a + b = |m - n| + 2\ell , \quad \ell \text{ an integer} \geq 1 . \tag{8.5}$$

The idea is to show that the irreducibility of the degenerate series persists at the crossing points. Then it follows by a continuity argument that the unitarity established by (8.3) at one end of the line (8.4) extends along the line to the other end at the edge of the fundamental rectangle.

Fix k and ℓ, and let us reparametrize the lines (8.4) and (8.5) about the crossing point by

$$(a,b) = (|m - n| + k + \ell + s , \ell - k + s) \quad \text{in the case of (8.4)}$$

$$(a,b) = (|m - n| + k + \ell + t , \ell - k - t) \quad \text{in the case of (8.5)} .$$

We denote the respective full induced representations along these lines by $U_1^+(s)$ and $U_1^-(t)$, the sign referring to the slope of the line. For any admissible representation π, let $\Theta(\pi)$ denote the global character.

We treat only $m \geq n$. Use of the intertwining operators (including knowledge of decompositions in $SL(2,\mathbb{C})$ and the fact that the image of the Langlands intertwining operator is irreducible) implies in deleted neighborhoods of $s = 0$ and $t = 0$ that we have

$$\Theta(U_1^+(s)) = \Theta(D_1^+(s)) + \Theta(U_2(s)) \tag{8.6a}$$

$$\Theta(U_1^-(t)) = \Theta(D_1^-(t)) + \Theta(U_3(t)), \tag{8.6b}$$

with the characters on the right irreducible. Here $D_1^+(s)$ and $D_1^-(t)$ are degenerate series induced from finite-dimensional representations of a noncuspidal maximal parabolic subgroup, and $U_2(s)$ and $U_3(t)$ are induced from the minimal parabolic subgroup with data as follows:

$U_2(s)$: same σ_0 but $[m,n,a,b]$ replaced by $[m+k\,,\, n-k\,,\, m-n+\ell+s\,,\, \ell+s]$

$U_3(t)$: same σ_0 but $[m,n,a,b]$ replaced by $[m+\ell\,,\, n-\ell\,,\, m-n+k+t\,,\, k+t]$.

The decompositions (8.6) persist for $s = 0$ and $t = 0$, but the characters on the right may become reducible.

Similar analysis of the intertwining operators for $U_2(s)$ and $U_3(t)$ in neighborhoods of $s = 0$ and $t = 0$ (including 0 this time) shows that

$\Theta(U_2(s))$ is irreducible for $s \neq 0$

$\Theta(U_3(t))$ is irreducible for all t

$$\Theta(U_2(0)) = \Theta(D_2(0)) + \Theta(U_3(0)) \quad \text{irreducibly}. \tag{8.7}$$

Let $J(0)$ be the Langlands quotient of $U_1^+(0) = U_1^-(0)$.

We shall show shortly that the only irreducible composition factors that can occur in $U_1^+(0)$ are $J(0)$, $D_2(0)$, and $U_3(0)$, and we know $J(0)$ occurs with multiplicity one. Consideration of Gelfand-Kirillov dimension (see Lemma 2.3 of [16]) then shows that

$$\Theta(D_1^+(0)) = \Theta(J(0)) + u\Theta(D_2(0)) \tag{8.8}$$

for an integer $u \geq 0$. We are to show that $u = 0$.

Let ψ denote the effect on characters of tensoring with the finite-dimensional representation of $SU(N,2)$ with extreme weight ke_{N+2} and then projecting according to the infinitesimal character given by the sum of (8.1) and ke_{N+2}. Direct computation with the aid of Corollary 5.10 of Speh-Vogan [15] shows that

$$\begin{aligned} \psi\Theta(U_1^+(0)) = \psi\Theta(U_2(0)) &= \Theta(U_2'(0)) \\ \psi\Theta(U_3(0)) &= \Theta(U_3'(0)), \end{aligned} \tag{8.9}$$

where $U_2'(0)$ and $U_3'(0)$ are induced from the minimal parabolic subgroup with data as follows:

$U_2'(0)$: same σ_0 but $[m,n,a,b]$ replaced by $[m+k, n, m-n+\ell, \ell-k]$

$U_3'(0)$: same σ_0 but $[m,n,a,b]$ replaced by $[m+\ell, n-\ell+k, m-n+k, 0]$.

Just as in (8.7), we have

$$\Theta(U_2'(0)) = \Theta(D_2'(0)) + \Theta(U_3'(0)) \quad \text{irreducibly} \tag{8.10}$$

with $D_2'(0) \neq 0$. Applying ψ to (8.6a) at $s = 0$ and using (8.9), we see that $\psi\Theta(D_1^+(0)) = 0$. But then ψ applied to (8.8) shows that $u\Theta(D_2'(0)) = 0$ (and also $\psi\Theta(J(0)) = 0$). Since $D_2'(0) \neq 0$, we conclude $u = 0$.

We are left with showing that the only irreducible composition factors that can occur in $U_1^+(0)$ are $J(0)$, $D_2(0)$, and $U_3(0)$. If the infinitesimal character is integral at our crossing point, then the fact that the crossing point is inside the fundamental rectangle implies the infinitesimal character is orthogonal to four mutually orthogonal roots. Four is too many singularities for any representation attached to G or to the cuspidal maximal parabolic subgroup, and four implies that e_{N-1}, e_N, e_{N+1}, and e_{N+2} participate in the singularities in the case of the minimal parabolic subgroup. Then it follows that $J(0)$, $D_2(0)$, and $U_3(0)$ are the only irreducible representations with the same infinitesimal character as $U_1^+(0)$. When the infinitesimal character is not integral at the crossing point, then it is so far from being integral that it is not the infinitesimal character of any representation attached to G or the cuspidal maximal parabolic subgroup. Moreover, the only way it can be the infinitesimal character of a representation attached to the minimal parabolic is if the coefficients of e_{N-1}, e_N, e_{N+1}, and e_{N+2} are merely permuted among themselves (in which case we are led to $J(0)$, $D_2(0)$, and $U_3(0)$) or if $N \leq 6$. For $N \leq 4$ there are no crossing points under study, for $N = 5$ only σ trivial is of concern and it is handled by inspection of the integrality, and for $N = 6$, interchange of the first 4 entries of the infinitesimal character with the last 4 entries leads to a nontrivial change of the representation on the 8-element center. Proposition 8.1 follows.

Applying Theorems 6.1 and 2.1, we immediately obtain the following complementary result.

Proposition 8.2. For $SU(N,2)$ with the minimal parabolic, with σ such that the fundamental rectangle has $a_0 > 0$ and $b_0 > 0$, and with ν real, no points ν within the closed fundamental rectangle and the closed positive Weyl chamber correspond to unitary representations except those listed in (8.2), (8.3), and Proposition 8.1.

9. Series associated with cuspidal maximal parabolic

Let MAN be the cuspidal maximal parabolic subgroup of $SU(N,2)$, $N \geq 3$. Let $\mathfrak{b} \subseteq \mathfrak{k}$ be a compact Cartan subalgebra, let Δ be the roots of $(\mathfrak{g}^{\mathbb{C}}, \mathfrak{b}^{\mathbb{C}})$, and suppose that A is constructed by Cayley transform $\underset{\sim}{c}$ from a noncompact root α.

Proposition 9.1. Let σ be a discrete series representation of the M of the cuspidal maximal parabolic subgroup of $SU(N,2)$, $N \geq 3$, and let λ_0 be its infinitesimal character. Let $t_0 \geq 0$ be the least number such that $\lambda_0 + t_0\alpha$ is integral and fails to be orthogonal to at least one compact root and one noncompact root. For $t > 0$, $J(MAN, \sigma, t\underset{\sim}{c}(\alpha))$ is infinitesimally unitary for $0 < t \leq t_0$ and not otherwise.

Sketch of proof. Either $\lambda_0 + \frac{1}{2}\alpha$ is integral (the "tangent case") or λ_0 is integral (the "cotangent case"). We sketch the proof only in the tangent case.

Fix $(\Delta_-)^+$ to make λ_0 dominant for it. Replacing α by $-\alpha$ if necessary, we can arrange that $\lambda_0 + t_0\alpha$ is nonsingular with respect to the noncompact roots in Δ. Then we can introduce $\Delta^+ = (\Delta^+)_1$ as in §3 so that λ_0 is Δ^+ dominant and α is simple. Let $r = t_0 + \frac{1}{2}$. For $1 \leq j \leq r$, we shall introduce recursively positive systems $(\Delta^+)_j$, subsystems Σ_j generated by simple roots, and forms χ_j so that $((\Delta^+)_j, \Sigma_j, \chi_j)$ is a Zuckerman triple (see §7) with parameters corresponding to $J(MAN, \sigma, (j-\frac{1}{2})\underset{\sim}{c}(\alpha))$ and so that $((\Delta^+)_r, \Sigma_r, \chi_r)$ is nondegenerate. Theorem 7.1 then says that unitarity does not extend beyond $t_0\underset{\sim}{c}(\alpha)$.

In addition, each Σ_j will correspond to a subgroup of G of real rank one, Σ_{j+1} will be generated by Σ_j and the roots orthogonal to $\lambda_0 + (j - \frac{1}{2})\alpha$, and the members of $(\Delta^+)_j$ not in Σ_{j+1}

will remain in $(\Delta^+)_{j+1}$. Since $\lambda_0 + t\alpha$ can be orthogonal to roots only for t in $\mathbb{Z} + \frac{1}{2}$ and cannot be orthogonal to more than two distinct roots (up to sign) if $t > 0$, it follows that $\langle \lambda_0 + t\alpha, \beta \rangle$ is nonzero for $0 \leq t < t_0$ and β in Δ_r^+ but not Σ_r. Since $\langle \lambda_0, \beta \rangle > 0$, we see that

$$\langle \lambda_0 + t\alpha, \beta \rangle \geq 0 \quad \text{for } 0 \leq t \leq t_0 \text{ and } \beta \text{ in } \Delta_r^+ \text{ but not } \Sigma_r. \qquad (9.1)$$

Let L be the subgroup of G corresponding to Σ_r. Since L has real rank one, any nonunitary principal series representation of L that is orthogonal to two linearly independent roots is irreducible. Combining this fact, the inequality (9.1), and the theory of [15], we see that $U(MAN, \sigma, t\underset{\sim}{c}(\alpha))$ is irreducible for $0 \leq t < t_0$. Then a standard continuity argument shows that $J(MAN, \sigma, t\underset{\sim}{c}(\alpha))$ is unitary for $0 \leq t \leq t_0$.

Thus the whole issue is to construct $((\Delta^+)_j, \Sigma_j, \chi_j)$ and verify its properties. We define $\Sigma_1 = \{\pm\alpha\}$ and choose $(\Delta^+)_j$ to make $\lambda_0 + (j - \frac{1}{2} - \epsilon)\alpha$ dominant (for $\epsilon > 0$ small). The $(\Delta^+)_j$ positive roots that are orthogonal to $\lambda_0 + (j - \frac{1}{2})\alpha$ are $(\Delta^+)_j$ simple, and Σ_{j+1} is taken as the root system generated by Σ_j and these roots if $j < r$. We define χ_j by

$$\chi_j - \rho_{(\Delta^+)_j} + 2\rho_{\Sigma_j} = \lambda_0 + (j - \tfrac{1}{2})\alpha .$$

Then we can verify all the asserted properties.

The tricky step (and here we use real-rank$(G) \leq 2$), is to check recursively that the minimal K-type is given by

$$\Lambda_j = \chi_j - 2\rho_{(\Delta_K^+)_j} + 2\rho_{(\Sigma_K^+)_j} .$$

In passing from Σ_j to Σ_{j+1}, we adjoined two simple roots, one compact (say $\beta_{j,c}$) and one noncompact (say $\beta_{j,n}$). The fact that the real rank is ≤ 2 enters the proof of the identity

$$2\rho_{\Sigma_{j+1},n} = 2\rho_{\Sigma_j,n} - \beta_{j,n} + \beta_{j,c} + \alpha \quad \text{for } 1 \leq j < r,$$

which is proved at the same time as the fact that Σ_{j+1} corresponds to a group of real rank one. Then it follows easily that $\Lambda_j = \Lambda_{j+1}$ for $1 \leq j < r$, and one shows that Λ_1 is the minimal K-type by using Theorem 1 and Lemma 3 of [9]. As we remarked at the end of §7, Λ_r is necessarily $(\Delta_K^+)_r$ dominant.

10. Duflo's method and the basic cases for SU(N,2)

Duflo [3] succeeded in proving that certain nonunitary principal series representations in some complex groups are not unitary by computing some determinants associated to an intertwining operator explicitly and finding two K-types on whose sum the operator is indefinite. In [11] we indicated how Duflo's method can be adapted to real groups. In this section we shall apply the method to the basic cases in SU(N,2) in order to prove Theorem 2.1. We are indebted to P. Delorme for a useful suggestion that helped us in this analysis.

We have already established in §8, especially in Proposition 8.1, that the representations asserted to be unitary in Theorem 2.1a are indeed unitary. For the representations in part (d) of the theorem, the unitarity follows from the standard continuity argument for the intertwining operator, which is scalar at $\nu = 0$.

Also in §8 we noted that (c) of the theorem follows from results about minimal K-types. The claimed nonunitarity in (d) follows from the Dirac inequality (Lemma 7.2); Δ^+ is taken either from the standard ordering given by indices $(1,2,\ldots,N+2)$ or from the ordering $(N+1,N+1,1,2,\ldots,N)$. The remainder of Theorem 2.1 consists of assertions of nonunitarity that we shall prove with Duflo's method and supplementary applications of Lemma 7.2.

Sample detailed calculations appear in the case of $SU(2,2)$ in [10]. In dealing with $SU(N,2)$, we must replace $SL(2,\mathbb{R})$ in that kind of calculation by $SU(N-1,1)$. Thus we need to know the scalar value of an $SU(N-1,1)$ intertwining operator on a K-type in a nonunitary principal series representation. We shall give that information, then say what intertwining determinants arise from certain K-types of $SU(N,2)$, and finally tell what nonunitarity is established by each of the K-types. The actual calculations of the intertwining determinants, which are carried out in the style of [10], will be omitted.

In $G_1 = SU(N-1,1)$, write

$$M_1 = \begin{pmatrix} \omega & & \\ & e^{i\theta} & \\ & & e^{i\theta} \end{pmatrix},$$

and let $\sigma_1 \leftrightarrow e^{ip\theta}$ with $|p| \leq N-2$. The minimal K_1-type is one-dimensional, and we normalize the intertwining operator to act as 1 on it. The K_1-types in the induced representation are the ones of the form

$$\overset{\quad N-1 \quad 1}{\begin{pmatrix} u & \\ & e^{i\theta} \end{pmatrix}} \longrightarrow e^{iq\theta}\, \tau_{(k,0,\ldots,0,\ell)}(u), \tag{10.1}$$

where $k \geq 0 \geq \ell$ and $q = p-k-\ell$. Let the A_1 parameter be af_1, where $2f_1$ is the real root. Reformulating some identities of Klimyk and Gavrilik [7] suitably, we find that the intertwining operator is given on the K_1-type (10.1) by the scalar

$$\Big[\prod_{j=0}^{k-1}\Big(\frac{2j-(a+p-N+1)}{2j+(a-p+N-1)}\Big)\Big]\Big[\prod_{j=\ell+1}^{0}\Big(\frac{2j+(a-p-N+1)}{2j-(a+p+N-1)}\Big)\Big]. \qquad (10.2)$$

The denominators are nonvanishing for $a \geq 0$ and for our purposes can be discarded.

Returning to $SU(N,2)$, consider the basic case $\sigma \leftrightarrow e^{i(m\theta+n\varphi)}$ with $m \geq n$, and form the representations of K given by

$$\tau\begin{pmatrix}\alpha & 0\\ 0 & \beta\end{pmatrix}_{\!N\;2}(P\otimes Q)\begin{pmatrix}w_1\\ \\ w_N\end{pmatrix}\otimes\begin{pmatrix}z_1\\ z_2\end{pmatrix} = (\det\beta)^r P\Big(\alpha^{-1}\begin{pmatrix}w_1\\ \\ w_N\end{pmatrix}\Big)Q\Big(\beta^{-1}\begin{pmatrix}z_1\\ z_2\end{pmatrix}\Big).$$

We now give the functions of $\nu = af_1 + bf_2$ that arise as intertwining determinants from certain K-types. We neglect global constants and also irrelevant denominators like the ones in (10.2).

1) Fix an integer $\ell \geq 0$, let $r = m+\ell$, let $\{P\} = \mathbb{C}$, and let $\{Q\}$ be the holomorphic polynomials of degree $m-n+2\ell$. Then the intertwining determinant works out to be

$$\prod_{j=1}^{\ell}(a+b+n-m-2j)(a-b+n-m-2j).$$

2) Assume $m-n \geq 1$. Let $r = m$, let $\{P\}$ be the holomorphic polynomials of degree 1, and let $\{Q\}$ be the holomorphic polynomials of degree $m-n-1$. The intertwining determinant is

$$b-n-(N-1).$$

3) Assume $m-n \geq 1$. Let $r = m-1$, let $\{P\}$ be the antiholomorphic polynomials of degree 1, and let $\{Q\}$ be the holomorphic polynomials of degree $m-n-1$. The intertwining determinant is

$$a+m-(N-1).$$

4) Let $r = m - 1$, let $\{P\}$ be the antiholomorphic <u>alternating</u> tensors of rank 2, and let $\{Q\}$ be the holomorphic polynomials of degree $m - n$. The intertwining determinant is

$$[a + m - (N - 1)][b + n - (N - 1)] .$$

5) Let $r = m + 1$, let $\{P\}$ be the holomorphic <u>alternating</u> tensors of rank 2, and let $\{Q\}$ be the holomorphic polynomials of degree $m - n$. The intertwining determinant is

$$[a - m - (N - 1)][b - n - (N - 1)] .$$

We can use these determinants to exclude many representations from being unitary. If such a determinant has one sign in a region where unitary points occur, then no points are unitary in the region where the determinant takes on the opposite sign. From the determinants (1) it follows that no points with $a - b \neq m - n + 2k$ for an integer $k \geq 1$ are unitary except those in the triangles listed in Theorem 2.1a. If $m \geq 0$ and $m - n \geq 1$, then (2) and (3) exclude all points outside the fundamental rectangle if $n < -m$, and (3) suffices by itself if $n \geq -m$.

If $m = n$, then either (4) or (5) excludes points in the interior of the region to the right or above the fundamental rectangle (but not both), and one can exclude all the remaining points outside the fundamental rectangle except (2.3) by using a suitable Dirac inequality (Lemma 7.2). For the Dirac inequality one forms Δ^+ from the standard ordering $(1,2,\ldots,N+2)$ or from the ordering $(N+1,N+2,1,2,\ldots,N)$.

Finally if $0 > m > n$, then (2) excludes points strictly above the fundamental rectangle, and (5) excludes any other points to the right of the fundamental rectangle except those on the same horizontal as the top edge. One can then exclude all the remaining points outside the fundamental rectangle except (2.3) by using a suitable Dirac inequality (Lemma 7.2). For the Dirac inequality one forms Δ^+ from the standard ordering (1,2,...,N+2). This completes the proof of Theorem 2.1.

References

[1] M. W. Baldoni Silva, The unitary dual of Sp(n,1), $n \geq 2$, Duke Math. J. 48 (1981), 549-583.

[2] M. W. Baldoni Silva and D. Barbasch, The unitary spectrum for real rank one groups, preprint, 1982.

[3] M. Duflo, Représentations unitaires irréductibles des groupes simples complexes de rang deux, Bull. Soc. Math. France 107 (1979), 55-96.

[4] T. J. Enright, R. Howe, and N. R. Wallach, Unitarizable highest weight representations, Proceedings of conference at University of Utah 1982, to appear.

[5] A. Guillemonat, Sur l'unitarisation des modules sphériques: une extension de la bande critique, preprint, Université d'Aix-Marseille II, 1980.

[6] H. P. Jakobsen, Hermitian symmetric spaces and their unitary highest weight modules, preprint, 1981.

[7] A. U. Klimyk and A. M. Gavrilik, The representations of the groups U(n,1) and SO(n,1), preprint ITP-76-39E, Institute for Theoretical Physics, Kiev, USSR, 1976.

[8] A. W. Knapp, Investigations of unitary representations of semisimple Lie groups, Topics in Modern Harmonic Analysis, Istituto di Alta Matematica, to appear.

[9] A. W. Knapp, Minimal K-type formula, this volume.

[10] A. W. Knapp and B. Speh, Irreducible unitary representations of SU(2,2), J. Func. Anal. 45 (1982), 41-73.

[11] A. W. Knapp and B. Speh, Status of classification of irreducible unitary representations, "Harmonic Analysis Proceedings, Minneapolis 1981," Springer-Verlag Lecture Notes in Math. 908 (1982), 1-38.

[12] A. W. Knapp and E. M. Stein, Intertwining operators for semisimple groups, Ann. of Math. 93 (1971), 489-578.

[13] A. W. Knapp and G. J. Zuckerman, Classification of irreducible tempered representations of semisimple groups, Ann. of Math. 116 (1982).

[14] H. Schlichtkrull, The Langlands parameters of Flensted-Jensen's discrete series for semisimple symmetric spaces, preprint, 1981.

[15] B. Speh and D. A. Vogan, Reducibility of generalized principal series representations, Acta Math. 145 (1980), 227-299.

[16] D. A. Vogan, Gelfand-Kirillov dimension for Harish-Chandra modules, Inventiones Math. 48 (1978), 75-98.

[17] D. A. Vogan, The algebraic structure of the representation of semisimple Lie groups I, Ann. of Math. 109 (1979), 1-60.

[18] D. A. Vogan, "Representations of Real Reductive Groups," Birkhäuser, Boston, 1981.

[19] G. Zuckerman, Tensor products of finite and infinite dimensional representations of semisimple Lie groups, Ann. of Math. 106 (1977), 295-308.

Department of Mathematics
Cornell University
Ithaca, New York 14853, U.S.A.

ON THE EXISTENCE OF A GENERALIZED WEIL REPRESENTATION

Ronald L. Lipsman

1. Introduction

In [6], Duflo has shown how to parameterize the generic irreducible unitary representations of an arbitrary Lie group G by means of admissible co-adjoint orbits. The matter of obtaining explicit realizations of these representations by harmonic induction is taken up by the author in [13]. This can be achieved whenever invariant metric polarizations (satisfying several other technical conditions—see [13, §2]) can be found. Unfortunately, in the most general situation, such polarizations may not exit. But the following scenario is quite common. Invariant metric polarizations may fail to exist for a disconnected group G, yet they do exist for the identity component G^o of G. Such is the case for reductive groups and also for amenable groups. Thus the representations of G^o may be realized by harmonic induction —and then perhaps one can apply the Mackey group extension method to $G^o \triangleleft G$ in order to obtain explicit realizations of the representations of G. To be truly explicit this requires an explication of the—perhaps projective—extension of a representation π of G^o to its stability group G_π in G. If the representation π is associated to the orbit of an admissible linear functional $\varphi \in \mathfrak{g}^*$, the task is to compute explicitly a (projective) representation $\tilde{\pi}$ of G_φ (in the space of π) which satisfies

$$\tilde{\pi}(g_\varphi)\pi(g_\varphi^{-1}g^o g_\varphi) = \pi(g^o)\tilde{\pi}(g_\varphi), \quad g_\varphi \in G_\varphi, \quad g^o \in G^o.$$

When G is nilpotent, it is known that there exists a canonical 2-fold covering group $\tilde{G}_\varphi$ of G_φ and an ordinary representation $\tilde{\pi}$ of $\tilde{G}_\varphi$ in the space of π that intertwines [5]. $\tilde{\pi}$ is the Weil (or Segal-Shale-Weil or metaplectic or oscillator or whatever) representation when G^o is the Heisenberg group. In this paper we shall show that the same result holds on any Lie group (see Theorem 2.3). We think of the resulting representation $\tilde{\pi}$ as a generalized Weil representation. Its existence is another illustration of the following principle: Since the orbit method is valid for arbitrary Lie groups, thus phenomena, which are described on nilpotent groups by means of the orbital realization of the ingredients of harmonic analysis, should have analogs for general

Lie groups. A previous illustration of this principle is the elaborate theory of orbital integral formulas for global characters on general Lie groups (see e.g. [8], [10], [11], [14]).

There is a related idea to which the results of this paper have application. As already indicated, whenever there exist invariant metric polarizations, the irreducible representations may be obtained by harmonic induction from G_φ via those polarizations [13]. But suppose we can only find metric polarizations which are not invariant—that means (in the usual terminology) that $\mathrm{Ad}_{\mathfrak{e}/\mathfrak{d}}(\exp \mathfrak{d})$ is compact. (In fact, as I shall show in a future publication, such polarizations always exist.) Then one can harmonically induce from G_φ^o —see Andler [1] for the case of a real polarization. The conjecture is essentially that the commuting algebra of that representation is naturally isomorphic to the commuting algebra of $\mathrm{Ind}_{G_\varphi^o}^{G_\varphi} e^{i\varphi}$ (not quite; $\tilde{G}_\varphi$ must be brought in—see §4 for the precise statement). Then the irreducible unitary representations corresponding to φ are parameterized by the constituents of the latter induced representation. We shall verify this conjecture in certain cases by using the generalized Weil representation (Theorem 4.1 and Remark 4.2(iii)).

The main result of this paper—i.e. the existence of the generalized Weil representation—is Theorem 2.3. It is explained in §2 and proven in §3. The proof is by induction on dim G. As in both [6] and [13], that requires handling the reductive case separately, and then using a group extension procedure over the nilradical. In §4 we derive the main application. For groups $M \lhd G$ of the same dimension, we show how the representations of G given by Duflo's orbital parameterization can be realized explicitly in terms of the representations of M and the generalized Weil representations associated to them. The precise result is contained in Theorem 4.1.

2. Basic terminology and statement of the main result

Let G be a Lie group, G^o its identity component. As in [6] and [13], we consider the set $AP(G)$ of admissible, well-polarizable elements $\varphi \in \mathfrak{g}^*$. The meaning of these terms is as follows. Let G_φ be the stability group of φ in G, G_φ^o its identity component, $\mathfrak{g}_\varphi = \mathrm{LA}(G_\varphi)$, B_φ the symplectic form $B_\varphi(X,Y) = \varphi[X,Y]$, $\tilde{G}_\varphi$ the canonical 2-fold cover of G_φ (see [6], [13]), $p_\varphi : \tilde{G}_\varphi \to G_\varphi$ the covering map and $\tilde{G}_\varphi^o = p_\varphi^{-1}(G_\varphi^o)$. Then

φ admissible means: there exists a (unique) unitary character $\chi = \chi_\varphi$ of $\tilde{G}_\varphi^o$ such that $d\chi_\varphi = i\varphi|_{\mathfrak{g}_\varphi}$, $\chi_\varphi(1,-1) = -1$ (where $(1,\pm 1) = \mathrm{Ker}\, p_\varphi$); and

φ well-polarizable means: there exists a subalgebra $\mathfrak{h} \subseteq \mathfrak{g}_C$ which is maximal totally isotropic for B_φ, solvable, and satisfies the complex Pukanszky condition.

When φ is admissible, the set

$$\mathfrak{X}_G(\varphi) = \{\text{unitary representations } \tau \text{ of } \widetilde{G}_\varphi : \tau|_{\widetilde{G}^o_\varphi} = (\dim \tau)\chi_\varphi\}$$

is not empty.

<u>Definition 2.1</u>. (i) φ is called <u>full</u> if G_φ is connected; (ii) φ is called <u>D-integral</u> if there is a unitary <u>character</u> χ of $\widetilde{G}_\varphi$ such that $d\chi = i\varphi|_{\mathfrak{g}_\varphi}$ and $\chi(1,-1) = -1$.
Clearly if φ is admissible and full, then it's D-integral. Conversely, if φ is D-integral, then it's admissible and at least one of the elements of $\mathfrak{X}_G(\varphi)$ is one-dimensional.

<u>Definition 2.2</u>. $\mathfrak{X}^1_G(\varphi) = \{\tau \in \mathfrak{X}_G(\varphi) : \dim \tau = 1\}$.
To summarize: $\mathfrak{X}_G(\varphi) \neq \emptyset$ exactly when φ is admissible; $\mathfrak{X}^1_G(\varphi) \neq \emptyset$ precisely when φ is D-integral.

Now to every pair (φ,τ), $\varphi \in AP(G)$, $\tau \in \mathfrak{X}_G(\varphi)$, Duflo has associated a unitary representation class $T_{\varphi,\tau}$ satisfying certain fundamental properties ([6, Intro.], [13, Prop. 2.6]). If there exists a polarization $\mathfrak{h}$ for φ which is invariant, metric and satisfies the strong (real) Pukanszky condition, then we can define the harmonically induced representation $\pi(\varphi,\tau,\mathfrak{h},q)$, $q \geqq 0$ [13, §2]. The main result of [13] is that if $\mathfrak{h}$ is also endlessly admissible and satisfies the Satake condition (see [13, §3]), and the last little group is of Harish-Chandra class, then

$\pi(\varphi,\tau,\mathfrak{h},q) = 0$ unless $q = q(\varphi,\mathfrak{h})$ (see [13, §2] for the definition of $q(\varphi,\mathfrak{h})$)

and

$$\pi(\varphi,\tau,\mathfrak{h},q(\varphi,\mathfrak{h})) \in T_{\varphi,\tau}.$$

For convenience we say $\mathfrak{h}$ is a <u>harmonic</u> polarization if it satisfies all the preceding conditions. Also we set $\pi(\varphi,\tau,\mathfrak{h}) = \pi(\varphi,\tau,\mathfrak{h},q(\varphi,\mathfrak{h}))$.

Now suppose A is a group of automorphisms of G fixing a linear functional $\varphi \in AP(G)$ which is D-integral. Then A leaves G_φ and B_φ invariant; so we have a natural homomorphism $A \to Sp(\mathfrak{g}/\mathfrak{g}_\varphi, B_\varphi)$. Motivated by [5] and [6] we set

$$\widetilde{A} = (\widetilde{A})_{B_\varphi} = \{(a,m) : a \in A,\ m \in Mp(\mathfrak{g}/\mathfrak{g}_\varphi, B_\varphi)$$
$$a \text{ and } m \text{ have the same image in } Sp(\mathfrak{g}/\mathfrak{g}_\varphi, B_\varphi)\}$$

$$\cong \{(a,r) : a \in A,\ r : \Lambda(\mathfrak{z}/\mathfrak{z}_\varphi) \to \text{8th roots of unity}$$
$$r^2 = s_a^{-1},\ s_a \text{ as in [6, Ch. I]}\}.$$

Then $\widetilde{A} \xrightarrow{p} A$, $p(a,r) = a$, is a 2-fold cover and $\widetilde{A}$ acts on G by $\widetilde{a}\cdot g = a\cdot g$, $a = p(\widetilde{a})$. It's also easy to check that $a\cdot(g_\varphi,r) = (a\cdot g_\varphi, a\cdot r)$ defines an action of A on $\widetilde{G}_\varphi$. Then $\widetilde{A}$ also acts (via p), and we note that $\widetilde{a}\cdot(1,-1) = a\cdot(1,-1) = (1,-1)$.

Let $\chi \in \mathfrak{X}^1_G(\varphi)$ and suppose A fixes χ. Then A fixes the class $T_{\varphi,\chi}$. Let $\pi_{\varphi,\chi}$ be sone representation in the class. We have

$$\widetilde{a}\cdot\pi_{\varphi,\chi} = a\cdot\pi_{\varphi,\chi} \cong \pi_{\varphi,\chi}.$$

Of course there may be an obstruction to extending $\pi_{\varphi,\chi}$ as an ordinary representation to the semi-direct product $A\cdot G$. However, we shall show in this paper that—as in the nilpotent case—there is no obstruction to extending to $\widetilde{A}\cdot G$. That is the first part of our main result (see below). In order to explain the remaining two parts we need some additional preparation.

First let T be a closed subgroup of G_φ. We have the 2-fold cover $p_\varphi : \widetilde{G}_\varphi \to G_\varphi$. Thus we may set $\widetilde{T} = p_\varphi^{-1}(T)$. But T acts on G—by inner automorphism — and fixes φ. Thus $\widetilde{T}$ is also defined by the above construction (with $A = T$). It is evident that the two groups are identical, and the notation is consistent. We shall use this notation momentarily, but next suppose $\mathfrak{h}$ is a harmonic polarization for φ. We employ the stantard terminology: $\mathfrak{d} = \mathfrak{h} \cap \mathfrak{g}$, $\mathfrak{e} = (\mathfrak{h} + \bar{\mathfrak{h}}) \cap \mathfrak{g}$, $D = G_\varphi \exp \mathfrak{d}$, $E = G_\varphi \exp \mathfrak{e}$. In addition the polarization determines two charactes— $\nu = \nu_{\mathfrak{h}}$ of $\widetilde{G}_\varphi$ and $\mu = \mu_{\mathfrak{h}}$ of G_φ — which satisfy

$$\nu^2 = \det \mathrm{Ad}_{\mathfrak{h}/(\mathfrak{z}_\varphi)_c} \qquad \nu(1,-1) = -1$$

$$\mu = |\det \mathrm{Ad}_{\mathfrak{g}/\mathfrak{d}}|^{1/2} \qquad |\nu| = \nu^{-1}$$

(see [6, Ch. I], [13, §2]). Next suppose the harmonic polarization $\mathfrak{h}$ is invariant under A. The characters ν, μ are then defined on $\widetilde{A}$. Furthermore, we may take $\pi_{\varphi,\chi} = \pi(\varphi,\chi,\mathfrak{h})$. Then there is a <u>natural</u> action $\pi'(a) = \pi'(\varphi,\chi,\mathfrak{h})(a)$ on the space of $\pi(\varphi,\chi,\mathfrak{h})$ which satisfies

$$\pi'(a)\pi(\varphi,\chi,\mathfrak{h})(a^{-1}\cdot g) = \pi(\varphi,\chi,\mathfrak{h})(g)\pi'(a), \quad a \in A, \quad g \in G.$$

What is it? In order to describe it, we must recall the construction of the representation $\pi(\varphi,\chi,\mathfrak{h})$ [13]. We have the Duflo shift $\psi = \psi_{\mathfrak{h}} = \frac{1}{2}\,\mathrm{tr}\,\mathrm{ad}_{\mathfrak{h}/\mathfrak{d}_c}$. (Incidentally, ψ is pure imaginary on $\mathfrak{z}_\varphi$

and since $d(\mu\nu) = \psi$, we see why $|\mu\nu| = 1$.) There is a unique unitary character $\chi' = \chi'_\varphi$ of G_φ which satisfies $d\chi' = (i\varphi + \psi_{\mathfrak{h}})|_{\mathfrak{g}_\varphi}$ — it's defined by $\chi' = \chi\mu\nu$. We also write χ' for the unique unitary character of D which extends χ'_φ and satisfies $d\chi' = (\iota\varphi + \psi_{\mathfrak{h}})|_{\mathfrak{d}}$. χ' and $\mathfrak{h}$ determine a holomorphic line bundle $L_{\chi'}$ on E/D. Also the base has an E-invariant measure [3, p. 115], and since the polarization is metric also an E-invariant hermitian structure. The natural action of E on this bundle defines a unitary representation of E in the space $H_2^q(E/D, L_{\chi'})$ of square-integrable, harmonic, $L_{\chi'}$-valued $(0,q)$ forms. It's non-zero precisely for $q = q_o = q(\varphi,\mathfrak{h})$ [13]; call that representation σ. Then

$$\pi = \pi(\varphi,\chi,\mathfrak{h}) = \mathrm{Ind}_E^G \sigma.$$

Now by assumption A preserves the entire structure. Thus it acts naturally on $H_2^{q_o}(E/D, L_{\chi'})$, giving a unitary representation σ' of A that satisfies $\sigma'(a)\sigma(a^{-1}\cdot e) = \sigma(e)\sigma'(a)$, $a \in A$, $e \in E$. Finally A acts on the space of the induced representation, i.e. on the space of Borel functions f (from G to the space of σ), that satisfy $f(ge) = \sigma(e)^{-1}|\det \mathrm{Ad}_{\mathfrak{g}/\mathfrak{e}}(e)|^{1/2} f(g)$ and a square integrability condition, by

$$\pi'(a)f(g) = \sigma'(a)|\det \mathrm{Ad}_{\mathfrak{g}/\mathfrak{e}}(a)|^{-1/2} f(a^{-1}\cdot g), \quad a \in A, \quad g \in G.$$

The group G acts by left translation, and the equation $\pi'(a)\pi(a^{-1}\cdot g) = \pi(g)\pi'(a)$ is easily verified.

Here is the main theorem.

<u>Theorem 2.3</u>. Let G be any Lie group, $\varphi \in AP(G)$ a D-integral functional, $\chi \in \mathfrak{X}_G^1(\varphi)$, $\pi_{\varphi,\chi}$ a representation in the class $T_{\varphi,\chi}$, and A a group of automorphisms of G fixing φ and χ. Then there exists an ordinary representation $\tilde{\pi}$ of $\tilde{A}$ in the space of $\pi = \pi_{\varphi,\chi}$ such that:

(i) $\tilde{\pi}(\tilde{a})\pi(a^{-1}\cdot g) = \pi(g)\tilde{\pi}(\tilde{a})$, $\quad g \in G$, $a \in \tilde{A}$;

(ii) $\tilde{\pi}(\tilde{t}) = \chi(\tilde{t})^{-1}\pi(t)$, if $t \in T = (G_\varphi)_\chi$;

(iii) If $\pi = \pi(\varphi,\chi,\mathfrak{h})$ for some harmonic polarization $\mathfrak{h}$ of φ invariant under A, and if the image of A in $\mathrm{Aut}\,\mathfrak{g}_c$ lies in $\mathrm{Ad}\,\mathfrak{g}_c$, then

$$\tilde{\pi}(a) = \frac{\nu_{\mathfrak{h}}(\tilde{a})}{|\nu_{\mathfrak{h}}(a)|}\pi'(a) = (\mu_{\mathfrak{h}}\nu_{\mathfrak{h}})(\tilde{a})\pi'(a), \quad \tilde{a} \in \tilde{A}, \quad a = p(\tilde{a}) \in A.$$

We carry out the proof in the next section.

3. Proof of the main theorem

We begin with a lemma that will be used repeatedly in the ensuing argument.

Lemma 3.1. Let G be a Lie group and $H \subseteq G$ a closed subgroup. Let ω be a unitary representation of H and set $\pi = \mathrm{Ind}_H^G \omega$. Suppose A is a group of automorphisms preserving H and the class of ω —and so also the class of π. Let $\tilde{\omega}$ be an ordinary representation of A in the space H_ω of ω which intertwines, i.e. $\tilde{\omega}(a)\omega(a^{-1}\cdot h) = \omega(h)\tilde{\omega}(a)$, $h \in H$, $a \in A$. Then:

(i) There exists a natural action $\tilde{\pi}$ of A in the space of π which gives an ordinary representation of A that intertwines, i.e. $\tilde{\pi}(a)\pi(a^{-1}\cdot g) = \pi(g)\tilde{\pi}(a)$, $g \in G$, $a \in A$;

(ii) Let $\beta : B \to H$ be a homomorphism of a Lie group B into H and ρ a unitary character of B. Take $A = \beta(B)$ acting on G by inner automorphism and suppose we have the equation

$$\tilde{\omega}(\beta(b)) = \rho(b)\omega(\beta(b)), \quad b \in B.$$

Then it is also true that

$$\tilde{\pi}(\beta(b)) = \rho(b)\pi(\beta(b)), \quad b \in B.$$

Proof. The space of π consists of Borel functions $f : G \to H_\omega$ that satisfy

$$f(gh) = \omega(h)^{-1}\mu(h)f(g), \quad h \in H, \quad g \in G$$

$$\mu = |\det \mathrm{Ad}_{\mathfrak{g}/\mathfrak{h}}|^{1/2}$$

$$\int_{G/H} |f|^2 < \infty$$

(see [13, §2α], [5, II.1]). G acts by left translation. We define

$$\tilde{\pi}(a)f(g) = |\det \mathrm{Ad}_{\mathfrak{g}/\mathfrak{h}}(a)|^{-1/2}\tilde{\omega}(a)(f(a^{-1}\cdot g)), \quad a \in A, \quad g \in G.$$

Then it's a routine calculation to verify that $\tilde{\pi}$ fulfills the obligations spelled out in (i). Also it is convenient—and consistent—to denote

$$\mu(a) = |\det \mathrm{Ad}_{\mathfrak{g}/\mathfrak{h}}(a)|^{1/2},$$

so that

$$\tilde{\pi}(a)f(g) = \mu(a)^{-1}\,\tilde{\omega}(a)(f(a^{-1}\cdot g)).$$

In any event, (i) is done and it remains to establish (ii). Let $b \in B$, $a = \beta(b)$. Then

$$\begin{aligned}\tilde{\pi}(\beta(b))f(g) &= \mu(\beta(b))^{-1}\,\tilde{\omega}(\beta(b))(f(a^{-1}\cdot g))\\ &= \mu(\beta(b))^{-1}\rho(b)\omega(\beta(b))(f)a^{-1}ga))\\ &= \rho(b)\mu(a)^{-1}\omega(a)\omega(a)^{-1}\mu(a)f(a^{-1}g)\\ &= \rho(b)f(a^{-1}g) \;=\; \rho(b)\pi(a)f(g).\end{aligned}$$

q.e.d.

Now we turn to the proof of Theorem 2.3. We assume at first that G is reductive. In [6] the class $T_{\varphi,\chi}$ is specified by the choice of a particular representation $\pi_{\varphi,\chi}$ in the class. This is done in four stages of increasing generality:

(a) G connected semisimple, φ elliptic

(b) G connected reductive, φ standard (see [6, Ch. III] for the precise meaning of standard)

(c) G disconnected, φ standard

(d) the general case.

We shall deal with these cases one at a time.

(a) This case requires no special treatment. It can be subsumed under the next case, and so I go immediately to

(b) Let G be connected reductive, $\varphi \in AP(G)$ standard. In particular, G_φ is a maximal anisotropic Cartan subgroup of G. Then G_φ is connected, so φ is full and <u>a fortiori</u> D-integral. Let π_φ be the corresponding representation [6, Ch. III]. Now let A be a group of automorphisms of G fixing φ. It fixes the class T_φ of π_φ. Then Duflo [6, Lemma 6, p. 39] demonstrates the existence of a representation S_φ of $\tilde{A}$ which intertwines π_φ. S_φ obeys two further properties: an equivariance property with respect to $\mathfrak{n}$-homology ([6, Lemma III.6]), which we won't need; and the equation

$$S_\varphi(\tilde{t}) = \chi(\tilde{t})^{-1}\pi_\varphi(t), \quad t \in T = G_\varphi = G_\varphi^{o} \tag{3.1}$$

(see [6, Ch. III, eqn. (3)]). Take $\tilde{\pi} = S_\varphi$. By what we have already observed, parts (i) and (ii) of Theorem 2.3 are true. Part (iii) is also true and is a consequence of the results in [13, §4]. Note the additional hypothesis on $\mathrm{Ad}\,A$ is necessary to keep the group $A/Z_A(\mathfrak{g})\cdot G$ in the Harish-Chandra class as required by [13, Thm. 2.12].

(c) Now let G be disconnected, but suppose φ is still standard.

Part of our assumption is that φ is D-integral, so let $\chi \in \mathfrak{X}_G^1(\varphi)$. Then Duflo defines $\pi_{\varphi,\chi}$ by

$$\pi_{\varphi,\chi} = \mathrm{Ind}_{G_\varphi G^o}^{G} \chi \otimes S_\varphi \times \pi_\varphi.$$

Since χ and S_φ are both $-\mathrm{Id}$ on $(1,-1)$, and because of equation (3.1), $\pi_{\varphi,\chi}$ is well-defined. Now $\chi \otimes S_\varphi \times \pi_\varphi$ acts on the space of π_φ, a representation of the connected group G^o. The intertwining action of $\widetilde{A}$ on that space is already defined in (b). Then the action of $\widetilde{A}$ on the space of $\pi_{\varphi,\chi}$, the induced representation, is given by Lemma 3.1. Properties (i) and (ii) of Theorem 2.3 are clear from Lemma 3.1 and case (b). As for (iii), it is evident from the inclusion $E \subseteq G_\varphi G^o$, the definition of $\pi_{\varphi,\chi}$ and Lemma 3.1, that it is no loss of generality to assume $G = G_\varphi G^o$. But then

$$\pi_{\varphi,\chi} = \chi \otimes S_\varphi \times \pi_\varphi \cong \chi' \otimes \pi' \times \pi_\varphi. \qquad [13, §4]$$

That is

$$\widetilde{\pi} = S_\varphi = \chi^{-1}\chi'\pi' = \mu\nu\pi'.$$

(d) Finally we consider an arbitrary reductive Lie group G and a D-integral $\varphi \in AP(G)$. As in [6], [13], we decompose the Cartan subalgebra $\mathfrak{g}_\varphi = \mathfrak{t} + \mathfrak{a} + \mathfrak{z}$, where $\mathfrak{z} = \mathrm{Cent}\,\mathfrak{g}$, $\mathfrak{t}$ is the compact part of $\mathfrak{g}_\varphi \cap [\mathfrak{g},\mathfrak{g}]$, and $\mathfrak{a}$ is the split part. Set $\nu = \varphi|_{\mathfrak{a}}$, $L = \mathrm{Cent}_G \nu$, $\mathfrak{v} = \sum_{(\nu,\alpha)>0} \mathfrak{g}^\alpha$, $V = \exp \mathfrak{v}$ (see [6, III.6]). Then $Q = LV$ is a "parabolic subgroup" and $\xi = \varphi|_{\mathfrak{l}} \in AP(L)$ is standard. Also $L_\xi = G_\varphi$ and ξ is D-integral. Corresponding to any $\chi \in \mathfrak{X}_G^1(\varphi)$ there is a naturally defined unitary character λ of $\widetilde{L}_\xi$ —see [6, III.6]—and $\pi_{\varphi,\chi}$ is defined by

$$\pi_{\varphi,\chi} = \mathrm{Ind}_{LV}^{G} \pi_{\xi,\lambda} \times 1.$$

Let A be a group of automorphisms fixing φ and χ. The structure described in the last paragraph is canonically determined by φ, so A preserves all of it. The representation $\widetilde{\pi}_{\xi,\lambda}(\widetilde{a})$ is defined by case (c). Therefore we once again apply Lemma 3.1 to obtain $\widetilde{\pi}_{\varphi,\chi}(\widetilde{a})$. It is clear that properties (i) and (ii) of Theorem 2.3 are satisfied. We must only substantiate property (iii). Suppose that $\mathfrak{h}$ is a metric (equivalently cuspidal) polarization for φ [13, Lemma 4.2]. Thus $\mathfrak{e} = \mathfrak{p}$ is a cuspidal parabolic subalgebra. As explained in [13, §4], we may assume $\mathfrak{p} \subseteq \mathfrak{q} = \mathfrak{l} + \mathfrak{v}$. We have $\mathfrak{p} = \mathfrak{m} + \mathfrak{n} \subseteq \mathfrak{l} + \mathfrak{v}$.

Now both $\widetilde{\pi}_{\varphi,\chi}$ and $\pi'_{\varphi,\chi}$ are defined via the standard case by the induced representation technique. Thus the character by which they differ is the same as the character by which the representations $\widetilde{\pi}_{\xi,\lambda}$ and $\pi'_{\xi,\lambda}$ differ. Thus the proof comes down to showing that

$$\frac{r\nu_{\mathfrak{h}}(a,r)}{|\nu_{\mathfrak{h}}(a,r)|} = s\,\frac{\nu_{\mathfrak{h}\cap\mathfrak{l}_c}(a,s)}{|\nu_{\mathfrak{h}\cap\mathfrak{l}_c}(a,s)|}$$

for any $a \in A$, $r : \Lambda(\mathfrak{g}/\mathfrak{g}_\varphi) \to$ roots of unity, $s : \Lambda(\mathfrak{l}/\mathfrak{l}_\xi) \to$ roots of unity. But this is a straightforward consequence of [6, eqn. 27, p. 18].

The proof of Theorem 2.3 in case G is reductive is completed. This enables us to employ induction in order to prove the result in general. If $\dim G = 0$, then $\mathfrak{g} = \{0\}$, $\varphi = 0$, $\widetilde{G}_\varphi = G \times \mathbb{Z}_2$ and $T_{\varphi,\chi}(g) = \chi(g,1)$. If A acts on G, it fixes $\varphi = 0$, and $\widetilde{A} = A \times \mathbb{Z}_2$. Part of our assumption is that A also fixes χ. Then we define $\widetilde{\pi}(\widetilde{a}) = \widetilde{\pi}(a,\varepsilon) = \varepsilon$, $\varepsilon = \pm 1$. Regarding the three properties of Theorem 2.3, we observe that:

(i) is obvious;

(ii) is clear by the definition of $\widetilde{\pi}$; and

(iii) is satisfied vacuously.

Now in order to use induction, we apply the Mackey group extension procedure to the nilradical $N = \text{nilrad}\, G$. We may assume $\dim N > 0$, since otherwise G is reductive, and that case is already done. Now (starting with $\varphi \in AP(G)$, D-integral) we must recall some facts and constructions from [6], [13]. Let $\theta = \varphi|_{\mathfrak{n}} \in AP(N)$. We may assume $\theta \neq 0$; otherwise it is a question of the reductive group G/N. Then $\theta(\mathfrak{n}_\theta) \neq 0$ [2, p. 328] and we put $\mathfrak{q}_\theta = \text{Ker}\,\theta|_{\mathfrak{n}_\theta}$, $\Omega_\theta = \exp \mathfrak{q}_\theta$. We set $K = \widetilde{G}_\theta$. The map $q \to (q,1)$ injects Ω_θ into K, and we denote $G_1 = K/Q$, $\mathfrak{g}_1 = LA(G_1) = \mathfrak{g}_\theta/\mathfrak{q}_\theta$. If $\dim \mathfrak{g}_1 = \dim \mathfrak{g}$, then $\mathfrak{g}$ is reductive. Therefore, we may assume $\dim \mathfrak{g}_1 < \dim \mathfrak{g}$. This will enable us to invoke the induction hypothesis.

Next set $\xi = \varphi|_{\mathfrak{g}_\theta} \in AP(K)$. Let $\chi \in \mathfrak{X}^1_G(\varphi)$. Then naturally associated to χ is a character $\sigma \in \mathfrak{X}^1_K(\xi)$ [6, Ch. IV] so that ξ is also D-integral. Let $\pi_{\xi,\sigma}$ be a representation in the class $T_{\xi,\sigma}$. Duflo obtains the class $T_{\varphi,\chi}$ by setting

$$\pi_{\varphi,\chi} = \text{Ind}^G_{G_\theta N}\, \pi_{\xi,\sigma} \otimes \widetilde{\gamma} \times \gamma$$

where γ is a realization of $\gamma_\theta \in \hat{N}$ and the representation $\widetilde{\gamma}$ is the canonical extension to $\widetilde{G}_\theta$ defined in [5].

Now let A be a group of automorphisms of G fixing φ. N is characteristic, so it is preserved. Therefore θ is fixed, and so G_θ, Q_θ and ξ are also preserved. The hypothesis of Theorem 2.3 further demands that A fix χ; therefore σ is also fixed. Hence by the induction assumption, Theorem 2.3 is true for $\pi_{\xi,\sigma}$ —i.e. there exists $\tilde{\pi}_{\xi,\sigma}(\tilde{a})$, $\tilde{a} \in \tilde{A}$, acting in the space of $\pi_{\xi,\sigma}$, satisfying properties (i)-(iii). It is also true that there is a representation $\tilde{\gamma}$ of $\tilde{A}$ in the space of γ, which satisfies (i)-(iii) (see [5]). Therefore $\tilde{A}$ acts on the space of $\pi_{\xi,\sigma} \otimes \tilde{\gamma} \times \gamma$ by $\tilde{\pi}_{\xi,\sigma}(\tilde{a}) \otimes \tilde{\gamma}(\tilde{a})$. The action of $\tilde{A}$ on the space of $\pi_{\varphi,\chi}$ is thus determined by Lemma 3.1. It remains to demonstrate the three properties (i)-(iii).

Property (i) is immediate by the lemma. Property (ii) also follows from Lemma 3.1, but a little elaboration is in order. First we invoke the fact that $(G_\theta)_\xi = G_\varphi N_\theta$ [2, Prop. II.1.3]. Next we observe—of course, by Lemma 3.1—that it is no loss of generality to assume $G = G_\theta N$. Then

$$\tilde{\pi}_{\xi,\sigma}(\tilde{t}) = \sigma(\tilde{t})^{-1}\pi_{\xi,\sigma}(t), \quad t \in (K_\xi)_\sigma \qquad \text{by induction}$$

$$\tilde{\gamma}(\tilde{t}) = \chi_\theta(\tilde{t})^{-1}\gamma(t), \quad t \in \tilde{N}_\theta \qquad \text{by [5].}$$

Therefore

$$\pi_{\varphi,\chi}(\tilde{t}) = \sigma(\tilde{t})^{-1}\chi_\theta(\tilde{t})^{-1}\pi_{\varphi,\chi}(t)$$

$$= \chi(\tilde{t})^{-1}\pi_{\varphi,\chi}(t), \quad t \in (G_\varphi)_\chi$$

because σ and χ_θ match up on N_θ.

Finally we come to property (iii). Let $\mathfrak{h}$ be a harmonic polarization for φ. By [13], we have $\pi(\varphi,\chi,\mathfrak{h}) = \mathrm{Ind}_{G_\theta N}^{G}\, \pi(\xi,\sigma,\mathfrak{h} \cap \mathfrak{g}_\theta)_c) \otimes \tilde{\gamma} \times \gamma(\theta,\mathfrak{h} \cap \mathfrak{n}_c)$. It is also clear from [13, Lemma 2.1 and §3a] that π' can be defined via induction from $G_\theta N$ as well as E ([6, Eqn. (27), p. 18] is relevant again). Therefore using Lemma 3.1 a final time, we see that the result follows from the easily verified equation

$$r\nu_{\mathfrak{h}}(a,r) = r_1\nu_{\mathfrak{h}\cap(\mathfrak{g}_\theta)_c}(a,r_1)r_2\nu_{\mathfrak{h}\cap\mathfrak{n}_c}(a,r_2)$$

where $r_1 : \Lambda(\mathfrak{g}_\theta/(\mathfrak{g}_\theta)_\xi) \to$ roots of unity, $r_2 : \Lambda(\mathfrak{n}/\mathfrak{n}_\theta) \to$ roots of unity, $r : \Lambda(\mathfrak{g}/\mathfrak{g}_\varphi) \to$ roots of unity, and the number $r(l_1 + l_2)r_1(l_1)^{-1}r_2(l_2)^{-1}$ is independent of $l_1 \in \Lambda(\mathfrak{g}_\theta/\mathfrak{g}_\theta)_\xi)$, $l_2 \in \Lambda(\mathfrak{n}/\mathfrak{n}_\theta)$ —see [6, §4]. This completes the proof of Theorem 2.3.

<u>Remark 3.2</u>. The additional hypothesis $\mathrm{Ad}_{\mathfrak{g}_c} A \subseteq \mathrm{Ad}\, \mathfrak{g}_c$ is present

only because of the restriction to Harish-Chandra class reductive groups in [13, Thm. 2.12]. As I stated there, I believe the main results of [13] to be true without that hypothesis. Likewise I believe Theorem 2.3 is valid without the additional condition on $\mathrm{Ad}_{\mathfrak{g}_{C}} A$.

4. An application

We now wish to exploit the generalized Weil representations $\tilde{\pi}$ for the purpose indicated in the introduction. Namely, one of our main goals is to give an explicit realization of the representation classes $T_{\varphi,\tau}$, $\varphi \in AP(G)$, $\tau \in \mathfrak{X}_G(\varphi)$ via harmonic induction. To invoke [13] requires harmonic polarizations, which don't always exist. However, a common occurrence is that they exist for G°, but not for G. Thus we want to use the explicit realizations of the representations of G° and the corresponding generalized Weil representations to give realizations of the representations of G. Here is our main result in that direction.

<u>Theorem 4.1</u>. Let M be a closed normal subgroup of the Lie group G. Suppose $\dim M = \dim G$. Let $\varphi \in AP(G) = AP(M)$. Suppose that as an element of $AP(M)$, φ is full—i.e. M_{φ} is connected. Let $\pi = \pi_{\varphi}$ be a representation in the class $T_{\varphi} \in \hat{M}$. Then for any $\tau \in \mathfrak{X}_G(\varphi)$,

$$\mathrm{Ind}_{G_{\varphi}M}^{G} \tau \otimes \tilde{\pi} \times \pi \in T_{\varphi,\tau}, \tag{4.1}$$

where $\tilde{\pi}$ is the generalized Weil representation of $\tilde{G}_{\varphi}$ in the space of π given by Theorem 2.3.

<u>Remarks 4.2</u>. (i) The product in formula (4.1) makes sense since both τ and $\tilde{\pi}$ are representations of $\tilde{G}_{\varphi}$ which take the value -Id on (1,-1). Also in $G_{\varphi} \cap M = M_{\varphi} = G_{\varphi}^{\circ}$ we have

$$(\tau \otimes \tilde{\pi})|_{M_{\varphi}} = (\dim \tau)\pi$$

because

$$\tau|_{\tilde{M}_{\varphi}} = (\dim \tau)\chi_{\varphi}$$

and

$$\tilde{\pi}|_{\tilde{M}_{\varphi}} = \chi_{\varphi}^{-1}(\pi_{\varphi} \circ p_{\varphi}) \qquad \text{(by Theorem 3.1(ii))}.$$

(ii) Theorem 4.1 is <u>not</u> the most general result one could state along these lines. Namely, one could allow $\varphi \in AP(M)$ to be D-integral instead of full. (In fact, one could even omit that assumption—but then various non-order 2 obstructions may enter due to extensions $\tau_1 \in \mathfrak{X}_M(\varphi)$). We still have a generalized Weil representation $\widetilde{\pi}$ corresponding to $\pi_{\varphi,\chi}$, $\chi \in \mathfrak{X}^1_G(\varphi)$. However, the result in that case is more complicated, and I feel that for the purposes of this paper, Theorem 4.1 is as far as I wish to go.

(iii) Theorem 4.1 justifies the remark in the introduction about the Duflo-Andler conjecture. Let G be a Lie group, $\varphi \in AP(G)$. Let $M = G^o$ and assume M_φ is connected. Suppose $\mathfrak{h}$ is a harmonic polarization for $\varphi \in AP(M)$, not necessarily invariant under G_φ. Then one can harmonically induce via $\mathfrak{h}$ from G^o_φ —i.e. extend χ'_φ to D^o, take harmonic forms up to E^o and then induce to G. That representation—denote it ${}^o\pi_G(\varphi,\mathfrak{h})$ —is obviously equivalent to $\mathrm{Ind}_M^G \pi_M(\varphi,\mathfrak{h})$. But it follows from Theorem 2.3 that

$$(4.2)\quad {}^o\pi_G(\varphi,\mathfrak{h}) = \mathrm{Ind}_M^G \pi_M(\varphi,\mathfrak{h}) \cong \mathrm{Ind}_{G_\varphi M}^G \mathrm{Ind}_M^{G_\varphi M} \pi_M(\varphi,\mathfrak{h}) \cong \mathrm{Ind}_{G_\varphi M}^G \rho \otimes \widetilde{\pi}_\varphi \times \pi_\varphi$$

where $\pi_\varphi = \pi_M(\varphi,\mathfrak{h}) \in \hat{M}$, $\widetilde{\pi}_\varphi$ is the generalized Weil representation of Theorem 2.3 and ρ is the quasi-regular representation

$$\rho = \mathrm{Ind}_{\widetilde{G}^o_\varphi}^{\widetilde{G}_\varphi} \chi_\varphi .$$

Suppose that ρ is type I. Then

$$\rho \cong \int^\oplus_{\mathfrak{X}_G(\varphi)} (\dim \tau)\tau d\tau .$$

By Theorem 4.1, we have

$$\begin{aligned} {}^o\pi_G(\varphi,\mathfrak{h}) &\cong \mathrm{Ind}_{G_\varphi M}^G \rho \otimes \widetilde{\pi}_\varphi \times \pi_\varphi \cong \int^\otimes_{\mathfrak{X}_G(\varphi)} (\dim \tau)\mathrm{Ind}_{G_\varphi M}^G \tau \otimes \widetilde{\pi}_\varphi \times \pi_\varphi d\tau \\ &\cong \int^\otimes_{\mathfrak{X}_G(\varphi)} (\dim \tau)\pi_{\varphi,\tau} d\tau . \end{aligned}$$

In particular the commuting algebras of ${}^o\pi_G$ and ρ are isomorphic. Even if ρ is not type I, one can use (4.2) to write down an isomorphism of the commuting algebras.

(iv) The proof of Theorem 4.1 is by induction on dim G. As usual we prove it separately for reductive groups and then use the group extension technique over the nilradical. Unfortunately the hypotheses are not quite hereditary for the little groups. Thus if N = nilrad G, $\mathfrak{n} = LA(N)$, $\theta = \varphi|_{\mathfrak{n}}$, $\xi = \varphi|_{\mathfrak{g}_\theta}$; although it is true that $(M_\theta)_\xi = M_\varphi N_\theta$ is connected, the relevant little group actually is $M_1 = \widetilde{M}_\theta/Q_\theta$

(see §3); and $(\widetilde{M}_\theta)_\xi = \widetilde{M}_\varphi N_\theta$ may not be connected. Thus we must prove a somewhat more general result than Theorem 4.1. In order to state that result we require some additional terminology.

Let G be a Lie group, Γ a closed subgroup of $Z_G = \text{Cent } G$, $\eta \in \hat{\Gamma}$ a unitary character. Then $\Gamma \subseteqq G_\varphi$ and $\gamma \to (\gamma,1)$ injects Γ into $\widetilde{G}_\varphi$.

Definition 4.3. (i) $\varphi \in AP(G)$ is called η-admissible if $\mathfrak{X}_{G,\Gamma}(\varphi,\eta) \neq \emptyset$ where (ii) $\mathfrak{X}_{G,\Gamma}(\varphi,\eta) = \{\varphi \in \mathfrak{X}_G(\varphi) : \tau|_\Gamma = (\dim \tau)\eta\}$.

Theorem 4.4. Let M and G be as in Theorem 4.1, $\varphi \in AP(G) = AP(M)$. Let $\Gamma \subseteqq M \cap Z_G$, $\eta \in \hat{\Gamma}$. Suppose that the set $\mathfrak{X}_{M,\Gamma}(\varphi,\eta)$ contains exactly one irreducible element and that it is a character χ. Let $\pi = \pi_\chi$ be a representation in the class $T_{\varphi,\chi} \in \hat{M}$. Then for any $\tau \in \mathfrak{X}_G(\varphi,\eta)$,

$$\text{Ind}_{G_\varphi M}^{G} \tau \otimes \widetilde{\pi} \times \pi \in T_{\varphi,\tau}.$$

Remarks 4.5. (i) The hypothesis on $\mathfrak{X}_M(\varphi,\eta)$ implies that $M_\varphi = M_\varphi^o\Gamma$.

(ii) Theorem 4.1 follows from Theorem 4.4 by taking $\Gamma = \{1\}$ in the latter.

Proof of Theorem 4.4. We prove the result first for reductive groups $M \triangleleft G$, $\mathfrak{m} = \mathfrak{g}$. We use the usual 4-stage method ([6, III.1), except that in this argument we can coalesce the first three. Thus we begin with the assumption:

$\varphi \in AP(G) = AP(M)$ is standard. Of course $M^o = G^o$. Let π_φ be the corresponding representation of $M^o = G^o$. Then since $M_\varphi = M_\varphi^o\Gamma$ we have

$$\pi_\chi = \pi_{\varphi,\eta} = \text{Ind}_{\Gamma M^o}^{M} \eta \times \pi_\varphi$$

(see [6, p. 42] or observe that $M_{\pi_\varphi} = M^oM_\varphi = M^o\Gamma$). We already saw in §3 that

$$\pi_{\varphi,\tau} = \text{Ind}_{G_\varphi G^o}^{G} \tau \otimes \widetilde{\pi}_\varphi \times \pi_\varphi.$$

Therefore

$$\pi_{\varphi,\tau} = \text{Ind}_{G_\varphi G^o}^{G} \tau \otimes \widetilde{\pi}_\varphi \times \pi_\varphi \cong \text{Ind}_{G_\varphi M}^{G} \text{Ind}_{G_\varphi \Gamma M^o}^{G_\varphi M} \tau \otimes \widetilde{\pi}_\varphi \times \pi_\varphi$$

$$\cong \ \mathrm{Ind}_{G_\varphi M}^{G}(\tau \otimes "\tilde{\pi}_\varphi" \times \mathrm{Ind}_{\Gamma M^o}^{M}(\eta \times \pi_\varphi)) \ \cong \ \mathrm{Ind}_{G_\varphi M}^{G} \tau \otimes \tilde{\pi}_\chi \times \pi_\chi .$$

The meaning of "$\tilde{\pi}_\varphi$" is exactly that inherent in Lemma 3.1—namely lifting the G_φ-action from the space of π_φ to the space of the induced representation. But that is precisely $\tilde{\pi}_\chi$.

Now we go to the general reductive situation. We have $M \triangleleft G$, $\mathfrak{m} = \mathfrak{g}$, and we set $\nu = \varphi|_{\mathfrak{a}}$ as in §3. We take $L = \mathrm{Cent}_G \nu$, $L \cap M = \mathrm{Cent}_M \nu$ so that $\xi = \varphi|_{\mathfrak{l}}$ is standard. Then

$$G_\varphi \ = \ L_\xi \quad \text{and} \quad M_\varphi \ = \ (L \cap M)_\xi \ = \ M_\varphi^o \Gamma \ = \ L_\xi^o \Gamma .$$

Furthermore

$$\pi_{\varphi,\tau} \ = \ \mathrm{Ind}_{LV}^{G} \pi_{\xi,\sigma} \times 1 \qquad \pi_\chi \ = \ \mathrm{Ind}_{(L\cap M)V}^{M} \pi_{\xi,\eta} \times 1 .$$

We apply the previous case to obtain

$$\pi_{\xi,\lambda} \cong \mathrm{Ind}_{L_\xi (L\cap M)}^{L} \sigma \otimes \tilde{\pi}_{\xi,\eta} \times \pi_{\xi,\eta} .$$

Thus

$$\pi_{\xi,\sigma} \times 1 \ \cong \ \mathrm{Ind}_{L_\xi (L\cap M)V}^{LV} \sigma \otimes \tilde{\pi}_{\xi,\eta} \times \pi_{\xi,\eta} \times 1$$

and

$$\begin{aligned} \pi_{\varphi,\tau} \ &= \ \mathrm{Ind}_{LV}^{G} \pi_{\xi,\sigma} \times 1 \\ &\cong \ \mathrm{Ind}_{LV}^{G} \mathrm{Ind}_{L_\xi (L\cap M)V}^{LV} \sigma \otimes \tilde{\pi}_{\xi,\eta} \times \pi_{\xi,\eta} \times 1 \\ &\cong \ \mathrm{Ind}_{L_\xi (L\cap M)V}^{G} \sigma \otimes \tilde{\pi}_{\xi,\eta} \times \pi_{\xi,\eta} \times 1 \\ &\cong \ \mathrm{Ind}_{G_\varphi M}^{G} \mathrm{Ind}_{L_\xi (L\cap M)V}^{G_\varphi M} \sigma \otimes \tilde{\pi}_{\xi,\eta} \times \pi_{\xi,\eta} \times 1 \\ &\cong \ \mathrm{Ind}_{G_\varphi M}^{G} \tau \otimes \tilde{\pi}_\chi \times \mathrm{Ind}_{(L\cap M)V}^{M} \pi_{\xi,\eta} \times 1 \\ &\cong \ \mathrm{Ind}_{G_\varphi M}^{G} \tau \otimes \tilde{\pi}_\chi \times \pi_\chi . \end{aligned}$$

In the preceding computation we have invoked "induction in stages" twice, and omitted a rather easy verification of the next to last equivalence. That completes the proof in case $\mathfrak{g}$ is reductive.

Now for the induction portion of the argument. If $\dim G = 0$, then $\mathfrak{g} = \{0\}$, $\varphi = 0$, $G_\varphi = G$ and $M_\varphi = M$. The hypothesis forces

$M = \Gamma$. Then $\pi_\chi = \eta$, $\widetilde{\pi}_\chi = 1$ and the result is clear. Next take $N =$ nilrad G, $\mathfrak{n} = LA(N)$, $\theta = \varphi|_{\mathfrak{n}}$. If $\theta = 0$, we are back in the reductive case G/N. So we set up the usual structure: $\xi = \varphi|_{\mathfrak{g}_\theta}$, $\mathfrak{q}_\theta =$ Ker $\theta|_{\mathfrak{n}_\theta}$, $Q_\theta = \exp \mathfrak{q}_\theta$, $K = \widetilde{G}_\theta$, $G_1 = K/Q$. Then $\xi \in AP(G_1)$. Furtheremore there is a bijective correspondence $\tau \to \sigma$, $\mathfrak{X}_{G,\Gamma}(\varphi,\eta) \to \mathfrak{X}_{G_1,\Gamma_1}(\xi,\eta_1)$ [6, Lemma IV.6], where Γ_1 and η_1 are defined as follows. $\Gamma \subseteq M_\theta \subseteq G_\theta$ gives rise to $\widetilde{\Gamma} \subseteq \widetilde{M}_\theta \subseteq \widetilde{G}_\theta$. We set $\Gamma_1 = \widetilde{\Gamma}N_\theta/Q_\theta$, a central subgroup of G_1. (Note: Γ_1 is not trivial even if Γ is.) η_1 is the unique character of Γ_1 which is trivial on Q_θ, agrees with χ_θ on N_θ and obeys $\eta_1(\gamma,\pm 1) = \pm\eta(\gamma)$, $\gamma \in \Gamma$. χ_θ and η are compatible, so this is well-defined. Now of course

$$\pi_{\varphi,\tau} = \mathrm{Ind}_{G_\theta N}^{G} \pi_{\xi,\sigma} \otimes \widetilde{\gamma} \times \gamma, \quad \gamma = \gamma_\theta \in \hat{N}.$$

There is an analogous set-up for M:

$$M_1 = \widetilde{M}_\theta/Q_\theta; \quad \chi \to \lambda, \quad \mathfrak{X}_{M,\Gamma}(\varphi,\eta) \to \mathfrak{X}_{M_1,\Gamma_1}(\xi,\eta_1) \text{ is a bijection;}$$

$$\pi_\chi = \mathrm{Ind}_{M_\theta N}^{M} \pi_\lambda \otimes \widetilde{\gamma} \times \gamma.$$

The point is that the data G_1, M_1, ξ, Γ_1, η_1, σ, λ satisfy the same hypotheses as the original data. Well the fact that dim G_1 = dim M_1 is obvious. Also the fact that $\Gamma_1 \subseteq M_1 \cap \mathrm{Cent}\, G_1$ is immediate from the construction of Γ_1. Since the correspondence $\mathfrak{X}_{M,\Gamma}(\varphi,\eta) \leftrightarrow \mathfrak{X}_{M_1,\Gamma_1}(\xi,\eta_1)$ is bijective [6, Lemma IV.6], the latter contains only one element λ, and by its definition $\dim \lambda = \dim \chi$ — so its a character. Finally $\dim G_1 = \dim G$ can only occur if $\mathfrak{g}$ is reductive. Thus we are ready to use the induction hypothesis. It guarantees that

$$\pi_{\xi,\sigma} = \mathrm{Ind}_{(G_1)_\xi M_1}^{G_1} \sigma \otimes \widetilde{\pi}_\lambda \times \pi_\lambda.$$

But then we can compute

$$\begin{aligned}
\pi_{\varphi,\tau} &= \mathrm{Ind}_{G_\theta N}^{G} \pi_{\xi,\sigma} \otimes \widetilde{\gamma}|_{\widetilde{G}_\theta} \times \gamma \\
&= \mathrm{Ind}_{G_\theta N}^{G} (\mathrm{Ind}_{(G_1)_\xi M_1}^{G_1} \sigma \otimes \widetilde{\pi}_\lambda|_{(\widetilde{G})_\xi} \times \pi_\lambda) \otimes \widetilde{\gamma}|_{\widetilde{G}_\theta} \times \gamma \\
&\cong \mathrm{Ind}_{G_\theta N}^{G} (\mathrm{Ind}_{(\widetilde{G}_\theta)_\xi \widetilde{M}_\theta}^{\widetilde{G}_\theta} \sigma \otimes \widetilde{\pi}_\lambda|_{\widetilde{K}_\xi} \times \pi_\lambda) \otimes \widetilde{\gamma}|_{\widetilde{G}_\theta} \times \gamma
\end{aligned}$$

$$\cong \operatorname{Ind}_{G_\theta N}^{G} \operatorname{Ind}_{\tilde{G}_\varphi \tilde{M}_\theta N}^{\tilde{G}_\theta N} \{\tau \otimes (\tilde{\pi}_\lambda|_{\tilde{G}_\varphi} \times \tilde{\gamma}|_{\tilde{G}_\varphi}) \times (\pi_\lambda \otimes \tilde{\gamma}|_{\tilde{M}_\theta} \times \gamma)\}$$

$$\cong \operatorname{Ind}_{\tilde{G}_\varphi \tilde{M}_\theta N}^{G} \{ \qquad " \qquad \}$$

$$\cong \operatorname{Ind}_{G_\varphi M}^{G} \operatorname{Ind}_{\tilde{G}_\varphi \tilde{M}_\theta N}^{G_\varphi M} \{ \qquad " \qquad \}$$

$$\cong \operatorname{Ind}_{G_\varphi M}^{G} \tau \otimes \tilde{\pi}_\chi|_{\tilde{G}_\varphi} \times \pi_\chi .$$

Once again we have utilized induction in stages, and omitted some routine verifications of unitary equivalences.

That finishes the proof of Theorem 4.4, and so also of Theorem 4.1. We conclude the paper with several additional remarks.

Remarks 4.6. (i) Theorem 4.1 should be compared to [5, Thm 8.1] and [13, Thm. 3.3]. Theorem 4.1 is not a generalization of those results, but rather a companion to them. In all cases one is relating the orbital realization of a representation of G to its Mackey group extension realization over a normal subgroup—in [5] and [13] it is over the nilradical; here it is over an open subgroup. It's reasonable to speculate that all of these theorems are special cases of some more general result on any normal subgroup.

(ii) The main cases I have in mind in Theorem 4.1 or Theorem 4.4 are:

(a) G reductive, $M = G^{\circ}$. Invariant metric polarizations always exist for $\varphi \in \mathcal{AP}(M)$ and the generic representations of G may be realized in terms of the harmonically induced representations of M and the corresponding generalized Weil representations.

(b) G amenable, $M = G^{\circ}$. Again invariant metric (even positive) polarizations exist. The same comment as in (a) applies. In this case it would be enough to take M satisfying $[M,\mathrm{Rad}(M)] \subseteqq N$ (see [4], [12]).

(iii) It's a fact that there exist connected groups G (non-reductive and non-amenable) and $\varphi \in \mathcal{AP}(G)$ for which no invariant metric polarizations exist. In that case Theorem 4.1 is not applicable. One needs a lower-dimensional subgroup M —see remark (i).

(iv) At the risk of belaboring the point, I reiterate that the value of Theorem 4.1 is that it reduces the problem of realizing the representations $\pi_{\varphi,\tau}$ to the problem of realizing the representation $\tilde{\pi}_\chi$. That is solved for inner automorphisms (by Theorem 23(iii) and when there exist invariant harmonic polarizations (by Theorem 2.3(iii).

In general the problem remains to be settled. We can say somewhat more. Namely in some sense, only two examples occur. The first is the Weil-type representation $\tilde{\gamma}$. As we pass through the little groups (see [13, before Def. 3.2]) we encounter extensions of the same form. The process terminates in a reductive group; and then we have the representation S_φ of Duflo. The issue is how to put these together explicitly rather than recursively. In any event, it seems reasonable to call the representation $\tilde{\pi}_\varphi$ of Theorem 2.3 a generalized Weil representation. So at this point, one has only proven the existence of the generalzied Weil representation. One would also like to prove some kind of uniqueness and then to explicitly construct it. Uniqueness is problematic. $\tilde{\pi}(\tilde{a})$ can be modified by any character of $\tilde{A}$ without changing property (i) of Theorem 2.3. If G is a Heisenberg group, then $(\mathrm{Aut}\ G)_\varphi$ is semisimple, so no character can arise. For other G, $(\mathrm{Aut}\ G)_\varphi$ may support characters. Of course properties (ii) and (iii) of Theorem 2.3 pin down $\tilde{\pi}_\varphi$ a great deal; but they seem not to provide a unique characterization in the general case. Perhaps it's too much to hope for since $T_{\varphi,\tau}$ is only defined recursively. Nevertheless there should be a canonical choice. In the semisimple case, the representation S_φ is uniquely specified by an equivariance condition relating to a certain $\mathfrak{n}$-homology space [6, Ch. III]. Perhaps there is an analog of that phenomenon in greater generality. Lastly there is the matter of an explicit construction of the generalized Weil representation. This is now very well understood for nilpotent groups [9] and solvable groups [7]. Can this type of analysis be extended to non-solvable groups? It seems likely that such a theory would involve the famous intertwining integral operators that are present in the semisimple case.

(v) Finally we note that—as in the nilpotent case—it seems likely to be an interesting problem to study the decomposition into irreducible representations of $\tilde{\pi} \in \mathrm{Rep}(\tilde{A})$.

References

1. M. Andler, "Sur des représentations construites par la méthode des orbites," C. R. Acad. Sci. Paris, vol. 290 (1980), pp. 873-875.
2. L. Auslander & B. Kostant, "Polarization and unitary representations of solvable Lie groups," Invent. Math., vol. 14 (1971), pp. 255-354.
3. P. Bernat et al, Représentations des Groupes de Lie Résolubles, Dunod, Paris, 1972.
4. J. Charbonnel & M. Khalgui, "Polarisations pour un certain types des groupes de Lie," C. R. Acad. Sci. Paris, vol. 287 (1978), pp. 915-917.
5. M. Duflo, "Sur les extensions des représentations irréductibles des groupes de Lie nilpotents," Ann. Scient. Éc. Norm. Sup., vol. 5 (1972), pp. 71-120

6. M. Duflo, "Construction de représentations unitaires d'un groupe de Lie," CIME course, Cortona, 1980, preprint.
7. H. Fujiwara, G. Lion & B. Magneron, "Operateurs d'entrelacement et calcul d'obstruction sur des groupes de Lie resolubles," Lecture Notes in Math., no. 880, pp. 102-136.
8. M. Khalgui, "Sur les caratères des groupes de Lie a radical cocompact," Bull. Soc. Math. France, vol. 109 (1981), pp. 331-372; see also "Caractères des groupes de Lie," preprint.
9. G. Lion, "Integrales d'entrelacement sur des groupes de Lie nilpotents et indices de Maslov," Lecture Notes in Math., no. 587, pp. 160-176.
10. R. Lipsman, "Characters of Lie groups II: real polarizations and the orbital-integral character formula," J. D'Anal. Math., vol. 31 (1977), pp. 257-286.
11. R. Lipsman, "Orbit theory and harmonic analysis on Lie groups with co-compact nilradical," J. Math. Pures et Appl., vol. 59 (1980), pp. 337-374.
12. R. Lipsman, "Orbit theory and representations of Lie groups with co-compact radical, J. Math. Pures et Appl., vol. 60 (1982), pp. 17-39.
13. R. Lipsman, "Harmonic induction on Lie groups," revised preprint, 1982.
14. W. Rosmann, "Kirillov's character formula for reductive Lie groups," Invent. Math., vol. 48 (1978), pp. 207-220.

This research supported in part by NSF MCS-82-00706.

Department of Mathematics
University of Maryland
College Park, MD 20742

Solution of a conjecture of Langlands

Floyd L. Williams

Abstract. We present in this paper the solution of Langlands' conjecture on the multiplicity of an integrable discrete series representation in $L^2(\Gamma\backslash G)$. We show that the conjecture is true in fact for infinitely many non-integrable discrete classes.

1. Introduction. Let G be a connected non-compact linear semisimple Lie group and let Γ be a co-compact discrete subgroup of G. The regular representation of G on $L^2(\Gamma\backslash G)$ decomposes as a direct sum of irreducible unitary representations $\pi \in \hat{G}$ (= the unitary dual of G) where each π has a finite multiplicity $m_\pi(\Gamma)$. We shall assume that G has the rank of a maximal compact subgroup K of G. Then G admits $\pi \in \hat{G}$ with $L^2(G)$ matrix coefficients. If in fact $\pi \in \hat{G}$ has $L^1(G)$ matrix coefficients (i.e. π is an integrable discrete series representation of G) then in [6], [7] Langlands has worked out methods (based on results of Selberg [11], [12] and Harish-Chandra [1], [2]) for computing the multiplicity $m_\pi(\Gamma)$. In particular if 1 is the only elliptic element of Γ he has shown that

$$(1.1) \qquad m_\pi(\Gamma) = (\text{volume of } \Gamma\backslash G)\ (\text{formal degree of } \pi)$$

Equation (1.1) shows in particular that π actually occurs in $L^2(\Gamma\backslash G)$. In [4], [5] Hotta and Parthasarathy showed that with some restrictions on π, equation (1.1) holds in fact for infinitely many non-integrable discrete series representations π. These "restric-

Research supported by NSF Grant No. PRM 8205819.

tions" were recently removed by the author who thus obtained in [14] the most general multiplicity result possible for the discrete series. There, as in [4], [5], $m_\pi(\Gamma)$ was expressed as the (explicitly computable) index of a twisted Dirac operator on $\Gamma\backslash G/K$. This geometric interpretation of $m_\pi(\Gamma)$ is analogous to the original geometric interpretation proposed by Langlands. We recall, briefly, Langlands conjecture.

Let $H \subset K$ be a Cartan subgroup of G and fix a G-invariant holomorphic structure on G/H. Given a non-singular parameter by which the discrete series representation $\pi = \pi_\Lambda$ is determined (modulo the action of the Weyl group of (H,K)) let L_Λ be the corresponding holomorphic homogeneous line bundle over G/H, and for $X = \Gamma\backslash G/H$ let $H^q(X,L_\Lambda)$ be the space Γ-invariant, L_Λ-valued harmonic C^∞ forms of type (o,q) on G/H (relative to some G-invariant Hermitian metrics on G/H and L_Λ). Alternatively $H^q(X,L_\Lambda)$ is the qth-dimensional cohomology of the sheaf of local Γ-invariant holomorphic sections of L_Λ on pre-images of open sets in X. Suppose that

(1.2) Λ is sufficiently far away from Weyl chamber walls.

Then a theorem of Griffiths [3], [7] says that $H^q(X,L_\Lambda) = 0$ except for $q = q_\Lambda$, where q_Λ is a distinguished integer (cf. (2.12) below where $\Lambda = \lambda+\delta$) completely determined by Λ. In [7] Langlands conjectured that

$$\dim H^{q_\Lambda}(X,L_\Lambda) = m_{\pi_\Lambda}(\Gamma) \tag{1.3}$$

for π_Λ integrable. Schmid in [8] proved (1.3) for Λ subject to (1.2). Recently the author was able to by-pass condition (1.2) altogether and established (1.3) not only for all integrable π_Λ, but also for infinitely many non-integrable discrete series representations. Our result, which at the same times improves Griffiths' vanishing theorem for the cohomology spaces $H^q(X,L_\Lambda)$, is presented in Theorem 2.11 below. On the other hand Theorem 2.11 depends very much on strong results of Schmid in [8], [9], [10] .

Our method of proof is algebraic; in particular it does not involve the Selberg trace formula. In essence we show that the Hochschild-Serre spectral sequence for "n-cohomology" degenerates under a condition (see condition (2.9) below) considerably less restrictive than (1.2).

2. The complexifications of the Lie algebras g_o, k_o, h_o of G, K, H will be denoted by g, k, h , respectively. Let Δ be the set of non-zero roots of (g,h) and fix an arbitrary system of positive roots $\Delta^+ \subset \Delta$. If g_β is the root space of $\beta \in \Delta$ we set

$$n = \sum_{\alpha \in \Delta^+} g_{-\alpha} \tag{2.1}$$

The quotient G/H can be assigned a unique G-invariant complex structure such that n is the space of anti-holomorphic tangent vectors at the origin. Let L be the lattice of differentials of characters of H ; $\Delta \subset L \subset h_R^* \overset{\text{def.}}{=\!=}$ the linear functionals on h with real-valued restriction to $\sqrt{-1}\, h_o$. If $\lambda \in L$, λ is integral (since G is linear) and λ induces a holomorphic homogeneous line bundle L_λ over G/H . Let S_λ be the sheaf of germs of local Γ-invariant holomorphic sections of L_λ on the inverse images of open sets in $X = \Gamma\backslash G/H$ under the map $G/H \to X$ (as in section 1). Given $\pi \in \hat{G}$ let H_π denote the Hilbert space of π and also the space of K-finite vectors in H_π . Then the sheaf cohomology $H^q(X,S_\lambda)$ and the Lie algebra cohomology $H^q(n,H_\pi)$ are related by the following

<u>Theorem</u> 2.2.
$$H^q(X,S_\lambda) = \sum_{\substack{\pi \in \hat{G} \\ \pi(\Omega) = (\lambda,\lambda+2\delta)1}} m_\pi(\Gamma)\, H^q(n,H_\pi)_{-\lambda}$$

for $q \geq o$, $\lambda \in L$, where Ω = the Casimir operator of G , $2\delta = \sum_{\alpha \in \Delta^+} \alpha$ (,) = the Killing form of G , and where $H^q(n,H_\pi)_{-\lambda}$ is the subspace of vectors in $H^q(n,H_\pi)$ transforming according to the character $e^{-\lambda}$ of H .

<u>Proof</u>: By Lemma 6' of [8]

(2.3)
$$H^q(X,S_\lambda) = \sum_{\pi \in \hat{G}} m_\pi(\Gamma) H^q(\pi)_{-\lambda}$$

where $H^q(\pi)$ is the qth formal harmonic space of $\pi \in \hat{G}$ and $H^q(\pi)_{-\lambda}$ is the subspace of vectors which transform, under the action of H, according to the character $e^{-\lambda}$. But $H^q(\pi)$ is H-isomorphic to the n-cohomology space $H^q(n,H_\pi)$ by Theorem 3.1 of [10]. On the other hand if $\chi_{-\lambda-\delta}$ is Harish-Chandra's character of the center of the universal enveloping algebra of g then, by a theorem of Casselman and Osborne, $H^q(n,H_\pi)_{-\lambda}$ vanishes unless the infinitesimal character of π coincides with $\chi_{-\lambda-\delta}$. Thus Theorem 2.2 follows.

The Hochschild-Serre spectral sequence generated by the subalgebra $k \cap n$ of n can be used (as in [10]) to compute the n-cohomology $H^q(n,H_\pi)_{-\lambda}$ in Theorem 2.2. Its E_1 terms are given by

(2.4)
$$E_1^{rs} = H^s(k \cap n,\ H_\pi \otimes \Lambda^r\ (p/p \cap n))_{-\lambda}$$

where $g = k + p$ is a Cartan decomposition of g. Assume that $\lambda + \delta$

is regular and let

(2.5) $P^{(\lambda)} = \{\alpha \in \Delta \mid (\lambda + \delta, \alpha) > 0\}$

be the corresponding system of positive roots; let $2\delta^{(\lambda)} = \sum_{\alpha \in P^{(\lambda)}} \alpha$.

Let Δ_k, Δ_n denote the set of compact, non-compact roots, respectively, and let $\Delta_k^+ = \Delta^+ \cap \Delta_k$, $\Delta_n^+ = \Delta^+ \cap \Delta_n$, $2\delta_k = \sum_{\alpha \in \Delta_k^+} \alpha$, $2\delta_n = \sum_{\alpha \in \Delta_n^+} \alpha$.

Let w_o be the unique compact Weyl group element such that $(w_o(-\lambda-\delta), \Delta_k^+) > 0$ and define

(2.6) $\Lambda = w_o(-\lambda-\delta) - \delta.$

Define $P^{(\Lambda)}$ for Λ as in (2.5) and set $P_n^{(\Lambda)} = P^{(\Lambda)} \cap \Delta_n$, $P_n^{(\lambda)} = P^{(\lambda)} \cap \Delta_n$, $2\delta_n^{(\Lambda)} = \sum_{\alpha \in P_n^{(\Lambda)}} \alpha$, $2\delta_n^{(\lambda)} = \sum_{\alpha \in P_n^{(\lambda)}} \alpha$, and

(2.7) $Q_\lambda = \{\alpha \in \Delta_n^+ \mid (\lambda + \delta, \alpha) > 0\}$.

The cardinality of a set S will be denoted by $|S|$.

Theorem 2.8. Let $\lambda \in L$ such that $\lambda + \delta$ is regular. Assume that λ satisfies the condition

(2.9) $(\lambda + \delta - \delta^{(\lambda)}, \alpha) > 0$ for every α in $P_n^{(\lambda)}$;

note that $\lambda + \delta - \delta^{(\lambda)}$ is $P^{(\lambda)}$ - dominant. Let $\pi \in \hat{G}$ such that $\pi(\Omega) = (\lambda, \lambda + 2\delta)1$. Then $H^s(k \cap n, H_\pi \otimes \Lambda^r(p/p \cap n))_{-\lambda} = 0$ unless (i) $\pi|_K$ contains the irreducible K module V_ν with Δ_k^+-highest weight $\nu = \Lambda + \delta_n + \delta_n^{(\Lambda)}$ where Λ is given in (2.6), (ii) $s = |\{\alpha \in \Delta_k^+ | (\lambda + \delta, \alpha) < 0\}|$, and (iii) $r = |Q_\lambda|$ (see (2.7)).

Proof: Suppose $H^s(k \cap n, H_\pi \otimes \Lambda^r(p/p \cap n))_{-\lambda} \neq 0$. Then $H^s(k \cap n, V_\nu \otimes \Lambda^r(p/p \cap n))_{-\lambda} \neq 0$ for some K-type $V_\nu \subset \pi|_K$. Take the positive system $\tilde{\psi}$ in (4.2) of [10] to be $P^{(\Lambda)}$. Then by Lemma 4.9 of [10] there is a compact Weyl group element w_1 and a subset $Q_1 \subset P_n^{(\Lambda)}$ such that $-\lambda - \delta = w_1(\nu + \delta_k - \delta_n^{(\Lambda)} + \langle Q_1 \rangle)$, where we write $\langle Q_1 \rangle = \sum_{\alpha \varepsilon Q_1} \alpha$ for any $Q_1 \subset \Delta$; also $s \overset{a.}{=} |\{\alpha \varepsilon \Delta_k^+ | w_1 \alpha \varepsilon \Delta_k^+\}|$, and $\delta_n - w_1(\delta_n^{(\Lambda)} - \langle Q_1 \rangle) = \langle T \rangle$ for some $T \subset \Delta_n^+$ with $|T| = r$. We can write $w_o w_1(\delta_n^{(\Lambda)} - \langle Q_1 \rangle) = \delta_n^{(\Lambda)} - \langle Q \rangle$ for some $Q \subset P_n^{(\Lambda)}$, since the $\delta_n^{(\Lambda)} - \langle Q_1 \rangle$, $Q_1 \subset P_n^{(\Lambda)}$, are weights of the spin representation. Let $w = (w_o w_1)^{-1}$ so that by (2.6), $w_o^{-1}(\Lambda + \delta) = -\lambda - \delta$ $= w_1(\nu + \delta_k - \delta_n^{(\Lambda)} + \langle Q_1 \rangle) \Longrightarrow \nu \overset{b.}{=} w(\Lambda + \delta) + \delta_n^{(\Lambda)} - \langle Q_1 \rangle - \delta_k$ $= w(\Lambda + \delta + \delta_n^{(\Lambda)} - \langle Q \rangle) - \delta_k$. We have $(\Lambda + \delta - \delta^{(\Lambda)}, \alpha) > 0$ for every α in $P_n^{(\Lambda)}$ exactly by hypothesis (2.9). Moreover $\pi(\Omega) = (\lambda, \lambda + 2\delta)1 = (\Lambda, \Lambda + 2\delta)1$ and since $V_\nu \subset \pi|_K$ we can apply Corollary 2.7 of [14] to equation b. to deduce that Q must be the empty set! This forces $w = 1$, since $\nu + \delta_k$ is Δ_k - regular and $\nu + \delta_k$, $\Lambda + \delta + \delta_n^{(\Lambda)}$ are both Δ_k^+ - dominant, and proves (i) in the statement of Theorem 2.8. Now $w_1 = w_o^{-1}$ (since $w = 1$) and $\{\alpha \varepsilon \Delta_k^+ | (-\lambda - \delta, \alpha) > 0\} = w_o^{-1}\{\alpha \varepsilon \Delta_k^+ | w_o^{-1}\alpha \varepsilon \Delta_k^+\}$ (by definition of w_o). Hence by equation a. $s = |\{\alpha \varepsilon \Delta_k^+ | (\lambda + \delta, \alpha) < 0\}$, which proves (ii). Finally $Q_1 = \phi$ since $w = 1$ and $Q = \phi$; this implies $\langle T \rangle = \delta_n - w_o^{-1}\delta_n^{(\Lambda)} = \delta_n + \delta_n^{(\lambda)} = \langle Q_\lambda \rangle$ or $T = Q_\lambda$ (as is easily seen); i.e. $r = |T| = |Q_\lambda|$, which completes the proof of Theorem 2.8.

The argument just given shows that for λ, π satisfying the conditions of Theorem 2.8, $\pi|_K$ contains no K-type of the form $\nu = w(\Lambda + \delta + \delta_n^{(\Lambda)} - \langle Q \rangle) - \delta_k$ where w is a compact Weyl group element and $Q \subset P_n^{(\Lambda)}$ is a non-empty set. Therefore since $\pi|_K$ contains the K-type $\Lambda + \delta_n + \delta_n^{(\Lambda)}$ we can conclude by Schmid's Theorem on the lowest K-type [9] that π is necessarily the discrete series representation $\pi_{\Lambda+\delta}$ corresponding to the regular element $\Lambda + \delta$. That is,

Corollary 2.10. $H^s(k \cap n, H_\pi \otimes \Lambda^r(p/p \cap n))_{-\lambda} = 0$ in Theorem 2.8 unless in fact π is Harish-Chandra's discrete series representation [2] $\pi_{\Lambda+\delta} = \pi^*_{\lambda+\delta}$ (the * denoting the contragradient representation) with lowest K-type $\Lambda + \delta_n + \delta_n^{(\Lambda)}$.

Conversely if s,r satisfy (ii), (iii) in Theorem 2.8 one sees that $\dim H^s(k \cap n, H_{\pi_{\Lambda+\delta}} \otimes \Lambda^r(p/p \cap n))_{-\lambda} = 1$, since the lowest K-type $\Lambda + \delta_n + \delta_n^{(\Lambda)}$ occurs in $\pi_{\Lambda+\delta}|_K$ exactly once.

Theorem 2.11. (Solution of Langlands conjecture). For $\lambda \in L$ such that $\lambda + \delta$ is regular, let $\pi_{\lambda+\delta}$ be the corresponding Harish-Chandra discrete series representation. Let

$$(2.12)\qquad q_{\lambda+\delta} = |\{\alpha \in \Delta_k^+ | (\lambda + \delta, \alpha) < 0\}| + |\{\alpha \in \Delta_n^+ | (\lambda + \delta, \alpha) > 0\}| .$$

If λ satisfies condition (2.9) (which is automatically the case if $\pi_{\lambda+\delta}$ is integrable by Theorem 8.2 of [13]) then the sheaf cohomology groups $H^q(X,S_\lambda)$ vanish for $q \neq q_{\lambda+\delta}$. Moreover $\dim H^{q_{\lambda+\delta}}(X,S_\lambda) = m_{\pi_{\lambda+\delta}}(\Gamma)$.

Proof: By Theorem 2.8 and Corollary 2.10 the E_1 terms in (2.4) vanish unless $\pi = \pi_{\Lambda+\delta}$, $r = |Q_\lambda|$ and $s = s_o \overset{\text{def.}}{=} |\{\alpha \in \Delta_k^+ | (\lambda + \delta, \alpha) < 0\}|$. Thus only $\pi_{\Lambda+\delta} = \pi^*_{\lambda+\delta}$ can contribute to the sum in Theorem 2.2, and for every q, $H^q(n, H_{\pi_{\Lambda+\delta}})_{-\lambda} = E_1^{|Q_\lambda|, q-|Q_\lambda|} = 0$ unless $q - |Q_\lambda| = s_o$, i.e. unless $q = |Q_\lambda| + s_o \equiv q_{\lambda+\delta}$, which gives the vanishing statement in Theorem 2.11. Since $H^{q_{\lambda+\delta}}(n, H_{\pi_{\lambda+\delta}})_{-\lambda} = E_1^{|Q_\lambda|, s_o}$ has dimension one by the remarks following Corollary 2.10, Theorem 2.11 follows by application of Theorem 2.2 again, where one notes that $m_{\pi^*}(\Gamma) = m_{\pi}(\Gamma)$ for any $\pi \in \hat{G}$.

Condition (2.9), which replaces the unsatisfactory condition (1.2) for $\Lambda = \lambda + \delta$, is the most general condition possible for the validity of the multiplicity formula given in Theorem 2.11. Clearly the latter formula holds not only for integrable discrete series representations (as we have pointed out in the statement following (2.12)), but also for infinitely many non-integrable discrete classes.

Department of Mathematics
University of Massachusetts
Amherst, Massachusetts 01003
U.S.A.

References

1. Harish-Chandra, Representations of semisimple Lie groups VI, Amer. J. Math. 78 (1956), 1-41, 564-628, and 79 (1957), 87-120, 193-257, 733-760.

2. ________, Discrete series for semisimple Lie groups II, Acta Math. 116 (1966), 1-111.

3. P. Griffiths and W. Schmid, Locally homogeneous complex manifolds, Acta Math. 123 (1969), 253-302.

4. R. Hotta and R. Parthasarathy, A geometric meaning of the multiplicities of integrable discrete classes in $L^2(\Gamma\backslash G)$, Osaka J. Math. 10 (1973), 211-234.

5. ________, Multiplicity formulae for discrete series, Inventiones Mathematicae 26 (1974), 133-178.

6. R. Langlands, The dimension of spaces of automorphic forms, Amer. J. Math. 85 (1963), 99-125.

7. ________, Dimension of spaces of automorphic forms, Proc. Symposia in Pure Math. IX (1966), 253-257.

8. W. Schmid, On a conjecture of Langlands, Annals of Math. 93 (1971), 1-42.

9. ________, Some properties of square integrable representations of semisimple Lie groups, Annals of Math. 102 (1975), 535-564.

10. ________, L^2 - cohomology and the discrete series, Annals of Math. 103 (1976), 375-394.

11. A. Selberg, Harmonic analysis and discontinuous groups in weakly symmetric Riemannian spaces with applications to Dirichlet series, J. Indian Math. Soc. 20 (1956), 47-87.

12. ________, Seminars on analytic functions, Institute for Advanced Study, Princeton, vol. 2, 152-161.

13. P. Trombi and V. Varadarajan, Asymptotic behavior of eigenfunctions on a semisimple Lie group, Acta Math. 129 (1972), 237-280.

14. F. Williams, Discrete series multiplicities in $L^2(\Gamma\backslash G)$, to appear in Amer. J. Math.

Vol. 873: Constructive Mathematics, Proceedings, 1980. Edited by F. Richman. VII, 347 pages. 1981.

Vol. 874: Abelian Group Theory. Proceedings, 1981. Edited by R. Göbel and E. Walker. XXI, 447 pages. 1981.

Vol. 875: H. Zieschang, Finite Groups of Mapping Classes of Surfaces. VIII, 340 pages. 1981.

Vol. 876: J. P. Bickel, N. El Karoui and M. Yor. Ecole d'Eté de Probabilités de Saint-Flour IX – 1979. Edited by P. L. Hennequin. XI, 280 pages. 1981.

Vol. 877: J. Erven, B.-J. Falkowski, Low Order Cohomology and Applications. VI, 126 pages. 1981.

Vol. 878: Numerical Solution of Nonlinear Equations. Proceedings, 1980. Edited by E. L. Allgower, K. Glashoff, and H.-O. Peitgen. XIV, 440 pages. 1981.

Vol. 879: V. V. Sazonov, Normal Approximation – Some Recent Advances. VII, 105 pages. 1981.

Vol. 880: Non Commutative Harmonic Analysis and Lie Groups. Proceedings, 1980. Edited by J. Carmona and M. Vergne. IV, 553 pages. 1981.

Vol. 881: R. Lutz, M. Goze, Nonstandard Analysis. XIV, 261 pages. 1981.

Vol. 882: Integral Representations and Applications. Proceedings, 1980. Edited by K. Roggenkamp. XII, 479 pages. 1981.

Vol. 883: Cylindric Set Algebras. By L. Henkin, J. D. Monk, A. Tarski, H. Andréka, and I. Németi. VII, 323 pages. 1981.

Vol. 884: Combinatorial Mathematics VIII. Proceedings, 1980. Edited by K. L. McAvaney. XIII, 359 pages. 1981.

Vol. 885: Combinatorics and Graph Theory. Edited by S. B. Rao. Proceedings, 1980. VII, 500 pages. 1981.

Vol. 886: Fixed Point Theory. Proceedings, 1980. Edited by E. Fadell and G. Fournier. XII, 511 pages. 1981.

Vol. 887: F. van Oystaeyen, A. Verschoren, Non-commutative Algebraic Geometry, VI, 404 pages. 1981.

Vol. 888: Padé Approximation and its Applications. Proceedings, 1980. Edited by M. G. de Bruin and H. van Rossum. VI, 383 pages. 1981.

Vol. 889: J. Bourgain, New Classes of $\mathcal{L}^p$-Spaces. V, 143 pages. 1981.

Vol. 890: Model Theory and Arithmetic. Proceedings, 1979/80. Edited by C. Berline, K. McAloon, and J.-P. Ressayre. VI, 306 pages. 1981.

Vol. 891: Logic Symposia, Hakone, 1979, 1980. Proceedings, 1979, 1980. Edited by G. H. Müller, G. Takeuti, and T. Tugué. XI, 394 pages. 1981.

Vol. 892: H. Cajar, Billingsley Dimension in Probability Spaces. III, 106 pages. 1981.

Vol. 893: Geometries and Groups. Proceedings. Edited by M. Aigner and D. Jungnickel. X, 250 pages. 1981.

Vol. 894: Geometry Symposium. Utrecht 1980, Proceedings. Edited by E. Looijenga, D. Siersma, and F. Takens. V, 153 pages. 1981.

Vol. 895: J.A. Hillman, Alexander Ideals of Links. V, 178 pages. 1981.

Vol. 896: B. Angéniol, Familles de Cycles Algébriques – Schéma de Chow. VI, 140 pages. 1981.

Vol. 897: W. Buchholz, S. Feferman, W. Pohlers, W. Sieg, Iterated Inductive Definitions and Subsystems of Analysis: Recent Proof-Theoretical Studies. V, 383 pages. 1981.

Vol. 898: Dynamical Systems and Turbulence, Warwick, 1980. Proceedings. Edited by D. Rand and L.-S. Young. VI, 390 pages. 1981.

Vol. 899: Analytic Number Theory. Proceedings, 1980. Edited by M.I. Knopp. X, 478 pages. 1981.

Vol. 900: P. Deligne, J. S. Milne, A. Ogus, and K.-Y. Shih, Hodge Cycles, Motives, and Shimura Varieties. V, 414 pages. 1982.

Vol. 901: Séminaire Bourbaki vol. 1980/81 Exposés 561–578. III, 299 pages. 1981.

Vol. 902: F. Dumortier, P.R. Rodrigues, and R. Roussarie, Germs of Diffeomorphisms in the Plane. IV, 197 pages. 1981.

Vol. 903: Representations of Algebras. Proceedings, 1980. Edited by M. Auslander and E. Lluis. XV, 371 pages. 1981.

Vol. 904: K. Donner, Extension of Positive Operators and Korovkin Theorems. XII, 182 pages. 1982.

Vol. 905: Differential Geometric Methods in Mathematical Physics. Proceedings, 1980. Edited by H.-D. Doebner, S.J. Andersson, and H.R. Petry. VI, 309 pages. 1982.

Vol. 906: Séminaire de Théorie du Potentiel, Paris, No. 6. Proceedings. Edité par F. Hirsch et G. Mokobodzki. IV, 328 pages. 1982.

Vol. 907: P. Schenzel, Dualisierende Komplexe in der lokalen Algebra und Buchsbaum-Ringe. VII, 161 Seiten. 1982.

Vol. 908: Harmonic Analysis. Proceedings, 1981. Edited by F. Ricci and G. Weiss. V, 325 pages. 1982.

Vol. 909: Numerical Analysis. Proceedings, 1981. Edited by J.P. Hennart. VII, 247 pages. 1982.

Vol. 910: S.S. Abhyankar, Weighted Expansions for Canonical Desingularization. VII, 236 pages. 1982.

Vol. 911: O.G. Jørsboe, L. Mejlbro, The Carleson-Hunt Theorem on Fourier Series. IV, 123 pages. 1982.

Vol. 912: Numerical Analysis. Proceedings, 1981. Edited by G. A. Watson. XIII, 245 pages. 1982.

Vol. 913: O. Tammi, Extremum Problems for Bounded Univalent Functions II. VI, 168 pages. 1982.

Vol. 914: M. L. Warshauer, The Witt Group of Degree k Maps and Asymmetric Inner Product Spaces. IV, 269 pages. 1982.

Vol. 915: Categorical Aspects of Topology and Analysis. Proceedings, 1981. Edited by B. Banaschewski. XI, 385 pages. 1982.

Vol. 916: K.-U. Grusa, Zweidimensionale, interpolierende Lg-Splines und ihre Anwendungen. VIII, 238 Seiten. 1982.

Vol. 917: Brauer Groups in Ring Theory and Algebraic Geometry. Proceedings, 1981. Edited by F. van Oystaeyen and A. Verschoren. VIII, 300 pages. 1982.

Vol. 918: Z. Semadeni, Schauder Bases in Banach Spaces of Continuous Functions. V, 136 pages. 1982.

Vol. 919: Séminaire Pierre Lelong – Henri Skoda (Analyse) Années 1980/81 et Colloque de Wimereux, Mai 1981. Proceedings. Edité par P. Lelong et H. Skoda. VII, 383 pages. 1982.

Vol. 920: Séminaire de Probabilités XVI, 1980/81. Proceedings. Edité par J. Azéma et M. Yor. V, 622 pages. 1982.

Vol. 921: Séminaire de Probabilités XVI, 1980/81. Supplément: Géométrie Différentielle Stochastique. Proceedings. Edité par J. Azéma et M. Yor. III, 285 pages. 1982.

Vol. 922: B. Dacorogna, Weak Continuity and Weak Lower Semicontinuity of Non-Linear Functionals. V, 120 pages. 1982.

Vol. 923: Functional Analysis in Markov Processes. Proceedings, 1981. Edited by M. Fukushima. V, 307 pages. 1982.

Vol. 924: Séminaire d'Algèbre Paul Dubreil et Marie-Paule Malliavin. Proceedings, 1981. Edité par M.-P. Malliavin. V, 461 pages. 1982.

Vol. 925: The Riemann Problem, Complete Integrability and Arithmetic Applications. Proceedings, 1979-1980. Edited by D. Chudnovsky and G. Chudnovsky. VI, 373 pages. 1982.

Vol. 926: Geometric Techniques in Gauge Theories. Proceedings, 1981. Edited by R. Martini and E.M.de Jager. IX, 219 pages. 1982.

Vol. 927: Y. Z. Flicker, The Trace Formula and Base Change for GL (3). XII, 204 pages. 1982.

Vol. 928: Probability Measures on Groups. Proceedings 1981. Edited by H. Heyer. X, 477 pages. 1982.

Vol. 929: Ecole d'Eté de Probabilités de Saint-Flour X – 1980. Proceedings, 1980. Edited by P.L. Hennequin. X, 313 pages. 1982.

Vol. 930: P. Berthelot, L. Breen, et W. Messing, Théorie de Dieudonné Cristalline II. XI, 261 pages. 1982.

Vol. 931: D.M. Arnold, Finite Rank Torsion Free Abelian Groups and Rings. VII, 191 pages. 1982.

Vol. 932: Analytic Theory of Continued Fractions. Proceedings, 1981. Edited by W.B. Jones, W.J. Thron, and H. Waadeland. VI, 240 pages. 1982.

Vol. 933: Lie Algebras and Related Topics. Proceedings, 1981. Edited by D. Winter. VI, 236 pages. 1982.

Vol. 934: M. Sakai, Quadrature Domains. IV, 133 pages. 1982.

Vol. 935: R. Sot, Simple Morphisms in Algebraic Geometry. IV, 146 pages. 1982.

Vol. 936: S.M. Khaleelulla, Counterexamples in Topological Vector Spaces. XXI, 179 pages. 1982.

Vol. 937: E. Combet, Intégrales Exponentielles. VIII, 114 pages. 1982.

Vol. 938: Number Theory. Proceedings, 1981. Edited by K. Alladi. IX, 177 pages. 1982.

Vol. 939: Martingale Theory in Harmonic Analysis and Banach Spaces. Proceedings, 1981. Edited by J.-A. Chao and W.A. Woyczyński. VIII, 225 pages. 1982.

Vol. 940: S. Shelah, Proper Forcing. XXIX, 496 pages. 1982.

Vol. 941: A. Legrand, Homotopie des Espaces de Sections. VII, 132 pages. 1982.

Vol. 942: Theory and Applications of Singular Perturbations. Proceedings, 1981. Edited by W. Eckhaus and E.M. de Jager. V, 363 pages. 1982.

Vol. 943: V. Ancona, G. Tomassini, Modifications Analytiques. IV, 120 pages. 1982.

Vol. 944: Representations of Algebras. Workshop Proceedings, 1980. Edited by M. Auslander and E. Lluis. V, 258 pages. 1982.

Vol. 945: Measure Theory. Oberwolfach 1981, Proceedings. Edited by D. Kölzow and D. Maharam-Stone. XV, 431 pages. 1982.

Vol. 946: N. Spaltenstein, Classes Unipotentes et Sous-groupes de Borel. IX, 259 pages. 1982.

Vol. 947: Algebraic Threefolds. Proceedings, 1981. Edited by A. Conte. VII, 315 pages. 1982.

Vol. 948: Functional Analysis. Proceedings, 1981. Edited by D. Butković, H. Kraljević, and S. Kurepa. X, 239 pages. 1982.

Vol. 949: Harmonic Maps. Proceedings, 1980. Edited by R.J. Knill, M. Kalka and H.C.J. Sealey. V, 158 pages. 1982.

Vol. 950: Complex Analysis. Proceedings, 1980. Edited by J. Eells. IV, 428 pages. 1982.

Vol. 951: Advances in Non-Commutative Ring Theory. Proceedings, 1981. Edited by P.J. Fleury. V, 142 pages. 1982.

Vol. 952: Combinatorial Mathematics IX. Proceedings, 1981. Edited by E. Billington, S. Oates-Williams, and A.P. Street. XI, 443 pages. 1982.

Vol. 953: Iterative Solution of Nonlinear Systems of Equations. Proceedings, 1982. Edited by R. Ansorge, Th. Meis, and W. Törnig. VII, 202 pages. 1982.

Vol. 954: S.G. Pandit, S.G. Deo, Differential Systems Involving Impulses. VII, 102 pages. 1982.

Vol. 955: G. Gierz, Bundles of Topological Vector Spaces and Their Duality. IV, 296 pages. 1982.

Vol. 956: Group Actions and Vector Fields. Proceedings, 1981. Edited by J.B. Carrell. V, 144 pages. 1982.

Vol. 957: Differential Equations. Proceedings, 1981. Edited by D.G. de Figueiredo. VIII, 301 pages. 1982.

Vol. 958: F.R. Beyl, J. Tappe, Group Extensions, Representations, and the Schur Multiplicator. IV, 278 pages. 1982.

Vol. 959: Géométrie Algébrique Réelle et Formes Quadratiques, Proceedings, 1981. Edité par J.-L. Colliot-Thélène, M. Coste, L. Mahé, et M.-F. Roy. X, 458 pages. 1982.

Vol. 960: Multigrid Methods. Proceedings, 1981. Edited by W. Hackbusch and U. Trottenberg. VII, 652 pages. 1982.

Vol. 961: Algebraic Geometry. Proceedings, 1981. Edited by J.M. Aroca, R. Buchweitz, M. Giusti, and M. Merle. X, 500 pages. 1982.

Vol. 962: Category Theory. Proceedings, 1981. Edited by K.H. Kamps, D. Pumplün, and W. Tholen, XV, 322 pages. 1982.

Vol. 963: R. Nottrot, Optimal Processes on Manifolds. VI, 124 pages. 1982.

Vol. 964: Ordinary and Partial Differential Equations. Proceedings, 1982. Edited by W.N. Everitt and B.D. Sleeman. XVIII, 726 pages. 1982.

Vol. 965: Topics in Numerical Analysis. Proceedings, 1981. Edited by P.R. Turner. IX, 202 pages. 1982.

Vol. 966: Algebraic K-Theory. Proceedings, 1980, Part I. Edited by R.K. Dennis. VIII, 407 pages. 1982.

Vol. 967: Algebraic K-Theory. Proceedings, 1980. Part II. VIII, 409 pages. 1982.

Vol. 968: Numerical Integration of Differential Equations and Large Linear Systems. Proceedings, 1980. Edited by J. Hinze. VI, 412 pages. 1982.

Vol. 969: Combinatorial Theory. Proceedings, 1982. Edited by D. Jungnickel and K. Vedder. V, 326 pages. 1982.

Vol. 970: Twistor Geometry and Non-Linear Systems. Proceedings, 1980. Edited by H.-D. Doebner and T.D. Palev. V, 216 pages. 1982.

Vol. 971: Kleinian Groups and Related Topics. Proceedings, 1981. Edited by D.M. Gallo and R.M. Porter. V, 117 pages. 1983.

Vol. 972: Nonlinear Filtering and Stochastic Control. Proceedings, 1981. Edited by S.K. Mitter and A. Moro. VIII, 297 pages. 1983.

Vol. 973: Matrix Pencils. Proceedings, 1982. Edited by B. Kågström and A. Ruhe. XI, 293 pages. 1983.

Vol. 974: A. Draux, Polynômes Orthogonaux Formels – Applications. VI, 625 pages. 1983.

Vol. 975: Radical Banach Algebras and Automatic Continuity. Proceedings, 1981. Edited by J.M. Bachar, W.G. Bade, P.C. Curtis Jr., H.G. Dales and M.P. Thomas. VIII, 470 pages. 1983.

Vol. 976: X. Fernique, P.W. Millar, D.W. Stroock, M. Weber, Ecole d'Eté de Probabilités de Saint-Flour XI – 1981. Edited by P.L. Hennequin. XI, 465 pages. 1983.

Vol. 977: T. Parthasarathy, On Global Univalence Theorems. VIII, 106 pages. 1983.

Vol. 978: J. Ławrynowicz, J. Krzyż, Quasiconformal Mappings in the Plane. VI, 177 pages. 1983.

Vol. 979: Mathematical Theories of Optimization. Proceedings, 1981. Edited by J.P. Cecconi and T. Zolezzi. V, 268 pages. 1983.

Vol. 980: L. Breen. Fonctions thêta et théorème du cube. XIII, 115 pages. 1983.

Vol. 981: Value Distribution Theory. Proceedings, 1981. Edited by I. Laine and S. Rickman. VIII, 245 pages. 1983.

Vol. 982: Stability Problems for Stochastic Models. Proceedings, 1982. Edited by V.V. Kalashnikov and V.M. Zolotarev. XVII, 295 pages. 1983.